PRACTICAL APPLICATIONS OF RESIDUAL STRESS TECHNOLOGY

Proceedings of the
Third International Conference
15-17 May 1991
Indianapolis, Indiana, USA

EDITED BY
Clayton Ruud

Sponsored by
The Residual Stress Committee
of the
Highway/Off-Highway Vehicles Division
of
ASM International®

Published by
ASM International®
Materials Park, Ohio 44073-0002

ASM International® is a Society whose mission is to gather, process and disseminate technical information. ASM fosters the understanding and application of engineered materials and their research, design, reliable manufacture, use and economic and social benefits. This is accomplished via a unique global information-sharing network of interaction among members in forums and meetings, education programs, and through publications and electronic media.

Copyright © 1991
by
ASM International®
All Rights Reserved

No part of this book may be reproduced, stored in a retrieval system, or transmitted, in any form or by any means, electronic, mechanical photocopying, recording, or otherwise, without the prior written permission of the publisher. No warranties, express or implied, are given in connection with the accuracy or completeness of this publication and no responsibility can be taken for any claims that may arise.

Nothing contained in this book is to be construed as a grant of any right or manufacture, sale, or use in connection with anymethod, process, apparatus, product, or composition, whether or not covered by letters patent or registered trademark, nor as a defense against liability for the infringement of letters patent or registered trademark.

Library of Congress Catalog Card Number: 91-74086
ISBN: 0-87170-430-7
SAN: 204-7586

ASM International®
Materials Park, Ohio 44073-0002

Printed in the United States of America

ORGANIZING COMMITTEE

Clayton Ruud
Penn State University
University Park, PA
Conference Chairman

Robert Buennede
Caterpillar Tractor Co.
East Peoria, IL

Paul Prevey
Lambda Research, Inc.
Cincinnati, OH

Chet Grant
BOC Powertrain-GM Corp.
Flint, MI

William Young
Dana Corporation
Richmond, IN

Leonard Mordfin
National Institute of Standards
and Technology
Gaithersburg, MD

Preface

Residual stresses are those stresses present in a solid in the absence of external forces, excluding gravity. These stresses cause elastic strain and are distributed both on the micro and macro scales; *i.e.* microstresses are variations in stress across a polycrystalline grain or from grain to neighboring grain. Macrostresses are usually considered to be the statistical average residual stress spanning several tens to hundreds of polycrystalline grains. This book is concerned mainly with macrostress, which is the quantity that has usually been attempted to be measured by mechanical (including blind hole drilling), X-ray diffraction, magnetic, ultrasonic, *etc.* techniques.

Residual stresses, both macro and micro, are usually formed when a portion of a component or workpiece undergoes nonuniform, permanent dimensional change. This can be either plastic deformation, as in, for example, cold rolling of metals; or elastic deformation, as in solid-state phase transformations in metals and ceramics. Inhomogeneous plastic deformation may be induced by temperature gradients, as by welding, sintering or heat treating as well as by forming, pressing or material removal processes. Thus, virtually all manufacturing processes used to shape a material into a useful component will cause residual stresses, even processes such as plating or chemical and physical vapor deposition of films.

This book is the proceedings of the ASM International Practical Applications of Residual Stress Technology Conference. This meeting was the third in the series of topical conferences on residual stresses. The first was held in Chicago in 1980 and chaired by L.J. Van de Walle, the second was held in Cincinnati in 1987 and chaired by W.B. Young, and this latest was in Indianapolis and was chaired by C.O. Ruud.

The proceedings include papers on the effects of residual stress on materials properties, for example, fatigue life. Measurement methods and techniques are covered in a number of papers that describe applications of X-ray diffraction, Barkhausen noise, ultrasonic velocity, and neutron diffraction. The modeling and prediction of residual stresses are described in a group of papers that include applications to metal welding, forging, and forming, as well as ceramic matrix composites. Also, the nature of residual stresses developed by various manufacturing processes are described in papers concerned with grinding, peening, and cold forming.

The continuity of this series of conferences is due to the Residual Stress Committee of the Highway/Off-Highway Vehicles Division of ASM International. The viability and vitality of the third meeting in Indianapolis was due to the conference organizing committee which included:

Robert Buenneke, Caterpillar Tractor Company
Paul Prevey, Lambda Research
Chet Grant, BOC Powertrain-GM Corporation
William Young, Dana Corporation
Leonard Mordfin, National Institute of Standards and Technology

Also, the conference organizers are indebted to Lisa Hemeyer and Jan Daquila of ASM International for their work on producing the conference and proceedings.

Clayton O. Ruud
Department of Industrial and Management Systems Engineering
The Pennsylvania State University
Conference Chairman

TABLE OF CONTENTS

Effects on Material Properties

Effect of Residual Stress on Fatigue of Structural Alloys 1
   W.P. Koster, Metcut Research Associates, Inc., Cincinnati, Ohio

Effects of Residual Stresses on the Fracture Toughness of Zircaloy-2 Tubes 11
   H.P. Mohamadian and A.R. Mirshams, Southern University, Baton Rouge, Louisiana
   M.E. Cunningham, Pacific Northwest Laboratory, Richland, Washington

Stress and Machinability 19
   J. Tiberg, Swedish Institute of Production Engineering Research, Scadviken, Sweden

Effect of Residual Stresses on the Stress-Corrosion Cracking of Austenitic Stainless Steel Pipe Weldments 27
   J.C. Danke, University of Tennessee, Knoxville, Tennessee

Measurement Methods and Techniques I

Assessment of Component Condition from X-Ray Diffraction Data Employing the Sin^2-Psi Stress Measurement Technique 39
   E.B.S. Pardue and L.A. Lowery, Technology for Energy Corporation, Knoxville, Tennessee

Problems with Nondestructive Surface X-Ray Diffraction Residual Stress Measurement 47
   P.S. Prevey, Lambda Research, Inc., Cincinnati, Ohio

Evaluation of the Stress Distribution in Welded Steel by Measurement of the Barkhausen Noise Level 55
   K. Tiitto, American Stress Technologies, Inc., Pittsburgh, Pennsylvania
   A.S. Wojtas, W.J.P. Vink, and G. denOuden, Delft University of Technology, Delft, The Netherlands

Ultrasonic Measurements of Residual Stress in Railroad Wheels 61
   R.E. Schramm, A.V. Clark, D.V. Mitrakovic, and S.R. Schaps, National Institute of Standards and Technology, Boulder, Colorado

Measurement Methods and Techniques II

Residual Stresses, Great and Small 69
   G.T. Blake, Wiss, Janney, Elstner Associates, Inc., Northbrook, Illinois

The Use of X-Ray Diffraction to Determine the Triaxial Stress State in Cylindrical Specimens 77
   P.S. Prevey and P.W. Mason, Lambda Research, Inc., Cincinnati, Ohio

Neutron Diffraction Measurements of Residual Stress Near a Pin Hole in a Solid-Fuel Booster Rocket Casing 83
   J.H. Root, R.R. Hosbons, and T.M. Holden, Atomic Energy of Canada Ltd. Research, Chalk River, Ontario, Canada

Residual Stress Characterization in Technological Samples 87
    H.J. Prask and C.S. Choi, National Institute of Standards and Technology, Gaithersburg, Maryland

Modeling and Predictions

In-Process Control and Reduction of Residual Stresses and Distortion in Weldments 95
    K. Masubuchi, Massachusetts Institute of Technology, Cambridge, Massachusetts

Measurement and Prediction of Residual Elastic Strain Distributions in Stationary and Traveling Gas Tungsten Arc Welds 103
    K.W. Mahin and W.S. Winters, Sandia National Laboratories, Livermore, California
    T. Holden and J. Root, Atomic Energy of Canada Ltd. Research Chalk River, Ontario, Canada

Residual Stress Measurements at Skip Fillet Welds 111
    W.Y. Shen, P. Clayton, and M. Scholl, Oregon Graduate Institute, Beaverton, Oregon

Characterization of Residual Stresses in an Eccentric Swage Autofrettaged Thick-Walled Steel Cylinder 123
    S.L. Lee, G.P. O'Hara, V. Olmstead, and G. Capsimalis, Benet Laboratories U.S. Army Armament Research, Development, and Engineering Center, Watervliet, New York

Evaluation of Surface and Subsurface Stresses with Barkhausen Noise: A Numerical Approach 131
    P. Francino and K. Tiitto, American Stress Technologies, Inc., Pittsburgh, Pennsylvania

Application of a Unified Viscoplastic Model of Residual Stress to the Simulation of Autoclave Age Forming 137
    S. Foroudastan and G. Boshier, Textron Aerostructures, Nashville, Tennessee
    J. Peddieson, Tennessee Technological University, Cookeville, Tennessee

Modelling of Residual Stresses in Whisker-Reinforced Ceramic Matrix Composites 145
    Z. Li, Argonne National Laboratory, Argonne, Illinois
    R.C. Bradt, University of Nevada, Reno, Nevada

Manufacturing Processes

Residual Stresses in Fuel Channel Rolled Joints in CANDU PHWRS 153
    S. Venkatapathi, Atomic Energy of Canada Ltd. Research, Chalk River, Ontario, Canada
    T.A. Hunter, G.E. Canada Ltd., Peterborough, Ontario, Canada
    G.D. Moan, Atomic Energy of Canada Ltd. CANDU, Mississauga, Ontario, Canada

Detection of Subsurface Tensile Stress in an Aircraft Engine Mainshaft Bearing Using Barkhausen Noise 161
    W.P. Ogden, Split Ballbearing, Lebanon, New Hampshire

Practical Measurements of Distortion and Residual Stresses Made by Welding and Pneumatic Hammer Peening Warm Weld Beads 169
    R.W. Hinton, R.W. Hinton Associates, Center Valley, Pennsylvania

Intelligent Design Takes Advantage of Residual Stresses 175
    J.S. Eckersley, Metal Improvement Company, Belleville, Michigan
    T.J. Meister, Metal Improvement Company, Blue Ash, Ohio

Effect of Residual Stress on Fatigue of Structural Alloys

W.P. Koster
Metcut Research Associates, Inc.
Cincinnati, Ohio

ABSTRACT

The influence of surface residual stress on the fatigue behavior of iron-, nickel- and titanium-base alloys is documented. Machining/surface finishing methods including grinding, milling, peening as well as electrically and chemically assisted techniques are considered. Generally, tensile stresses reduce fatigue strength; compressive stresses produce the opposite effect. Magnitudes are quantified.

THE LAST SEVERAL GENERATIONS of mechanical and metallurgical engineers have dealt in some way with the interrelation between a material's surface condition and the resulting fatigue strength. From the early days of analyzing the failures of rotating and reciprocating machinery, the detriment of sharp corners, steps in shaft diameters and tooling marks has been well recognized. Over a century ago, these features were identified as stress risers which caused localized stress overload leading to premature failure under many operating conditions. These observations supported the concept that, barring gross internal defects, the surface condition of a component had the effect of limiting performance of the material in the same way as does the weakest link in a chain.

More recently, within the lifetimes of many of us, the finish level or surface roughness measurement has been considered to be very significant in relation to exhibited fatigue life. Papers and texts published in the first half of this century frequently discuss this subject; many show a direct quantitative correlation between surface roughness and fatigue. But from the present viewpoint, most of these discussions were flawed to some degree. There is often an apparent relation between surface roughness and fatigue, but, as we now understand, the reality is much more complex. In overview, a typical analysis of a half century ago plots fatigue strength versus surface roughness in a given material in which the various roughness levels were produced experimentally by different surface finishing methods -- grinding, turning with different tool radii, rough polishing, finish polishing, lapping, etc. Coincidentally, a reasonable correlation could indeed be shown among these data. And a degree of curve smoothing made it look even better. What was not realized, however, was that the surface residual stresses resulting from these different finishing methods varied widely and just happened to permit a degree of correlation. For example, rough grinding typically produces high tensile residual stresses; "fine" grinding, lower tensile stresses; polishing and sanding, moderate compression; lapping and polishing result in high (but shallow) compressive stresses, etc. In retrospect, the reality was that variations in residual stress produced in the surface by the finishing methods were driving the range of fatigue life observed, not the surface roughness itself.

The whole phenomenon of surface integrity, namely study of the relation between the mechanical properties exhibited by a material and the characteristics of the surface of a specimen or component was studied quite extensively in recent years. Much of this work is described and summarized in a series of Air Force Technical Reports [1,2,3,]. In this work, the focus was on determining the effect of operating parameters applied to a broad array of material removal and finishing processes - turning, milling, grinding, ECM, EDM, electropolishing, mechanical polishing, shot peening and bead peening, etc. on subsequent mechanical behavior of materials. The purpose of this body of work was basically twofold: (1) to identify specific areas of material/process sensitivity so that necessary controls could be applied to assure appropriate hardware quality and (2), to identify areas of indifference, or low sensitivity of a material to a particular process, wherein controls could be relaxed and cost savings potentially achieved. The effort covered a practical range of parameters for each of the processes studied. Results were measured by material performance: tensile strength and ductility, stress rupture and creep behavior, fatigue and stress corrosion resistance. Concomitant effects on microstructure, microhardness and residual stress were also observed. Many interactions were noted but clearly the predominant property affected during the course of these surface integrity investigations was the fatigue strength. And it was equally clear that the variable characteristic that could be most consistently related to fatigue behavior was the surface residual stress.

Based on what is now known and documented, several conclusions pertinent to the subject of this paper can be drawn:

(1) "Surface" residual stress is a predominant factor influencing the fatigue strength of metals and alloys. In this context, "surface" is defined as a surface layer or zone ranging up to .005-.010" deep, not the extreme outer fibers in isolation. In many cases, the residual stress in this surface zone is identified as the single most important factor influencing fatigue behavior after the inherent strength of the material itself is considered.

(2) The residual stress effect is most pronounced in the infinite life stress range -- endurance limit regime, 10^7 cycles + -- although significant effects of residual stress continue to be seen in the higher cyclic stress ranges (low cycle fatigue region) which yield lives of 10^3 - 10^5 cycles.

(3) The effect of residual surface stress on fatigue tends to diminish as test temperatures are increased significantly, presumably related to the relaxation of residual stress due to the thermal exposure.

(4) The effect of surface finish or roughness within reasonable ranges (up to 100-200 microinches AA) on fatigue strength is much less than has been traditionally accepted by the engineering community. Residual stress is a much more potent factor. Gross surface discontinuities are, however, clearly detrimental.

(5) Surface residual stress also affects the stress corrosion resistance of those materials/conditions sensitive to this phenomenon. Tensile residual stresses in the same direction as applied external stress significantly reduce the threshold of crack initiation in sensitive materials/environmental conditions.

(6) Complex surface conditions -- such as those produced by carburizing and subsequent case hardening, for example -- exhibit the same trend although the increase in fatigue strength observed (in this case) is probably attributable to both compressive residual stresses related to the carburizing/hardening plus concomitant phase changes as well as to the increase in the inherent strength of the higher carbon, higher hardness of the surface.

(7) Shot peening is particularly complex in its analysis and can be discussed in overview only in general terms. Peening generally produces residual surface compression; it also typically causes work hardening. Both tend to increase fatigue life. Peening, however, also causes microcracking, microtearing, the formation of small laps and seams as well as significant surface roughening in the "overpeened" condition. The net result is commonly an increase in fatigue strength with moderate peening but a degradation of properties if multiple or redundant peening cycles are applied. Undoubtedly, the enhancement due to the development of compressive residual stresses is a factor in the behavior of peened surfaces but is only one of the factors which are active.

(8) The preceding conclusions are all relatable to "engineering materials" which exhibit "reasonable" ductility. The effects noted probably diminish when low ductility and even brittle materials are considered. On the other hand fatigue is frequently not a limiting consideration in designing with these low ductility materials, hence this area may be primarily of academic interest.

To support these ideas, I have assembled a number of illustrations. Going back to the basic surface integrity research, I would like to focus first on a study on AISI 4340 which had been quenched and tempered to a hardness of 50 Rockwell C. This effort very clearly illustrates and verifies the relation between endurance limit and residual stress in the surface of the material [3]. This evaluation was made on flat fatigue specimens tested at room temperature in the fully-reversed cantilever bending mode. Fifteen different sets of grinding parameters were used to produce the 15 different groups of fatigue specimens involved. Within the grinding process, variations included different grinding wheel abrasives, grit size and bonding hardness as well as ranges of grinding speed, fluid, infeed and grinding wheel dressing procedure. A plot of the resulting fatigue strengths versus surface residual stresses is shown in Figure 1. The microstructure related to each of the 15 tests conditions was very similar. The basic structure was that of tempered martensite although occasional patches of untempered martensite were observed in one or two of the more aggressively ground sets of samples. The surface roughness of all 15 groups of specimens was the same for practical purposes, in the range of 40-55 microinches AA.

Residual stresses typically measured in ground surfaces are characterized in Figure 2. These data were obtained by classical surface dissection or layer removal techniques. [4] X-ray diffraction procedures currently in widespread use yield similar results. [5,6] It is particularly important to note that over the range of grinding parameters studied, the tendency is to observe residual stresses at the very surface which are low in magnitude, often close to zero. As one penetrates into the surface, the residual stress profile unfolds either into the negative region indicating a compressive residual stress or to the positive region indicating a tensile residual stress. Figure 2 shows residual stresses characteristic of the range of test conditions shown in Figure 1 for the same material. Typically, the non-aggressive grinding procedures using low wheel speeds, "soft" grinding wheels, etc., labeled in this Figure as "gentle" resulted in fairly low level compressive stresses having a negative peak on the order of .0005" beneath the surface. So-called conventional grinding, using the range of parameters normally prescribed for grinding hardened steels, results in fairly high level tensile stresses, again peaking less than .001" beneath the surface. Abusive grinding procedures which are quite aggressive result in peak stresses not much greater than that produced by conventional grinding but the depth of the peak is somewhat further from the surface. Note also, as can be seen in Figure 2, that the

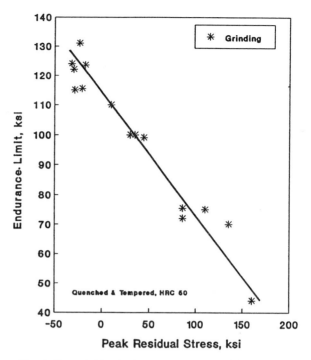

Fig. 1. Correlation between endurance limit and peak residual stress in AISI 4340 steel

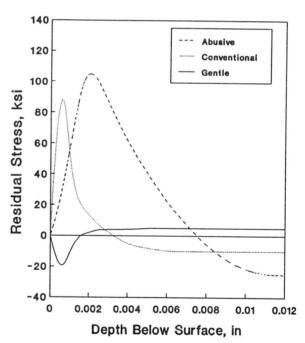

Fig. 2. Residual surface stress in AISI 4340 (Q & T, HRC 50) produced by surface grinding.

volume of surface material tensilely stressed as a result of abusive grinding is much greater than that similarly stressed by conventional grinding. This phenomenon frequently results in the distortion of components, especially thin ones, which are ground using aggressive procedures.

In analyzing this experimentation, several trial correlations were made between fatigue strength and other variations observed. The only meaningful correlation, and one which has been observed many times with other materials and other surface finishing parameters, was that with the "peak" stress observed, be it in tension or compression, as shown in Figure 2. Such "peak" values were the basis for charting the fatigue data shown in Figure 1. In rationalizing this behavior it was suspected that the low-life specimens, those which were known to contain high tensile residual stresses beneath the surface, would exhibit sub-surface failure initiation at the depth where the peak tensile residual stress was observed. Further, high-life specimens with compressively stressed subsurfaces would probably show fatigue cracking initiating at the surface. While this area of investigation was not pursued during the original work on the subject, the phenomenon has been subsequently observed in several instances. This is not a surprising observation, but merely supports the behavior to be expected from the nature of the residual stress distribution.

The observed relationship between residual stress in the surface zone and fatigue strength as measured by the 10^7 cycle endurance limit is shown for 17-4PH, a typical austenitic steel, in Figure 3; for Titanium 6Al-4V in Figure 4; for Rene' 41, a typical nickel base alloy, in Figure 5; and for Al 7075, typical of high strength aluminum alloys, in Figure 6. While Figure 1 involves only grinding, Figures 3 thru 6 show data from surface finishing methods including grinding, sanding, milling, turning, and non-traditional methods such as electrical discharge machining, electrochemical machining, electropolishing, chem-milling, etc. In Figure 3, two of the 17-4PH specimens exhibited distinct double peaks in the residual stress profile, as indicated in this figure. Not a perfect correlation, but certainly a reasonable trend of correlation is exhibited by all of these materials under the test conditions discussed. In the case of the titanium alloy shown in Figure 4, two different chemical milling procedures (non-traditional) were used which resulted in

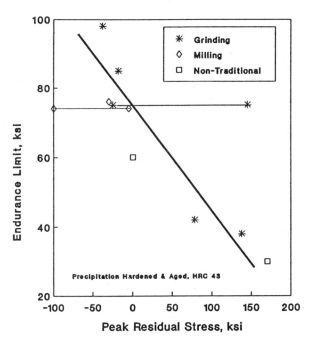

Fig. 3. Correlation between endurance limit and peak residual stress in 17-4 PH stainless steel

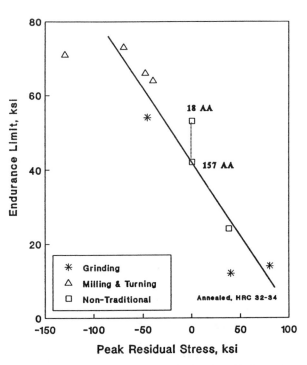

Fig. 4. Correlation between endurance limit and peak residual stress in Ti-6Al-4V

widely varying surface finishes; both exhibited virtually no residual stress. The surface finish is identified for these two specimens in Figure 4 (data points are linked) and indicates the magnitude of influence that can probably be attributed to surface roughness in this particular situation. Similar observation can be made with the data on Rene' 41 shown in Figure 5. An electrochemically machined group of specimens, having no residual stress, exhibited an endurance limit of 41 ksi and had a surface roughness of 17 microinches AA. A chemically milled sample, which also contained no residual stress had a surface roughness in excess of 200 microinches AA along with a fatigue strength of 33 ksi. (Data points again linked.) Presumably, this difference in fatigue is again an indication of the magnitude of the influence of surface roughness levels when they can be isolated in the absence of residual stress effects.

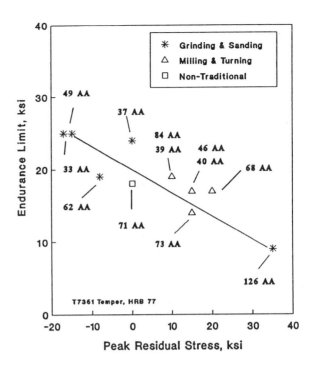

Fig. 6. Correlation between endurance limit and peak residual stress in Al 7075.

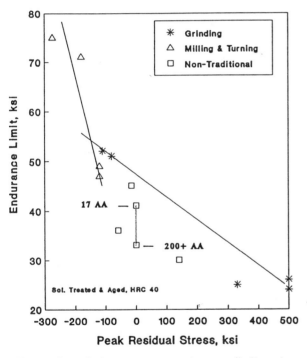

Fig. 5. Correlation between endurance limit and peak residual stress in Rene' 41.

Aluminum alloys tend to exhibit smaller differences in fatigue strength levels but then of course the absolute strength level is also low in these materials compared to the others under discussion. In Figure 6, surface roughness of the various specimens have been indicated to permit further contemplation over the effect, or relative lack thereof, of surface finish on the influence of fatigue behavior.

While the fatigue relationship shown in Figures 1 and 3 thru 6 are typical, they are not universal. Figure 7 summarizes a grinding parameter study on a nickel base alloy, Rene' 41, similar experimentally to the study on hardened steel (AISI 4340) synopsized in Figure 1. In the case of this material, the continuous degradation in fatigue strength with increasing tensile residual stresses is not observed. However, the enhancement associated with residual compressive stress is evident.

Figure 8 suggests a method of analyzing and utilizing residual stress information that may be more useful in exploring and controlling production processes. This chart summarizes data taken on experimentally produced samples of actual production hardware, in this case an inboard flap track fitting from the Boeing 747 airplane. Samples of flap track hardware were made from the normal production alloy, a modification of AISI 4340 hardened to 53 Rc, which were produced using various grinding parameters and subsequently tested to determine fatigue behavior. [3] In this case, the residual stress value used was not from the arbitrarily located "peak" from a complete residual stress profile but rather was, in all cases, the corrected value determined at a uniform depth of 0.0005" beneath the hardware surface. Using this data, a very

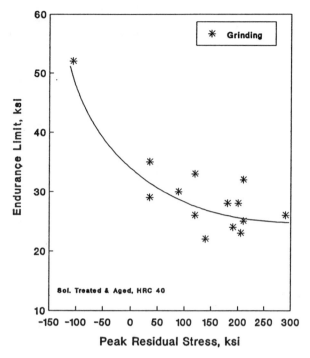

Fig. 7. Correlation between endurance limit and peak residual stress in Rene' 41.

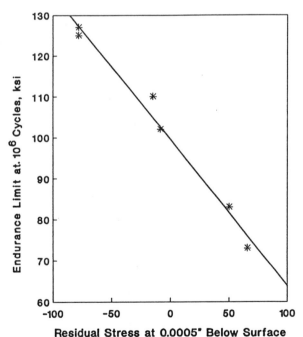

Fig. 8. Relationship between fatigue performance and residual stress in 4340 Mod. steel 747 Inboard Flaptrack Fittings.

good correlation between residual stress and fatigue was obtained as shown in Figure 8. The important thing to remember, and I feel it cannot be over emphasized, is that the residual stress which controls fatigue behavior is frequently sub-surface, not the outer fiber stress which is determined by a single non-destructive reading. Accordingly, serious residual stress measurement is an invasive or destructive procedure and in the investigation phase at least, can only be done on a sampling basis. On the other hand, once a process has been well characterized, surface non-destructive stress measurements may be useful for quality assurance purposes.

Because of the traditional importance attributed to surface roughness in relation to fatigue, it seems appropriate to present the data shown in Figure 9. This chart summarizes experimental work in which several different materials were studied using surface finishing methods intended to produce a range of surface roughness but with virtually identical residual stress distribution in each group. Referring to the top group of bars showing data for longitudinally ground 4340, grinding parameters were adjusted to produce surface finishes ranging from 8 microinches AA to 127 microinches AA. All of these samples exhibited peak residual stress of approximately 20-30 ksi in compression. Note that in this group the endurance limit ranged from 100 ksi for the roughest surface to 117 ksi for the smoothest surface. Since all pertinent variables were held as closely as possible and since the residual stresses were verified as being virtually identical, this magnitude of difference in fatigue strength, 100 to 117 ksi is believed attributable to the effect of surface roughness. Similar comments can be made for each of the groups of three bars shown in this chart. The third group, for abusively ground 4340, shows the depression in fatigue strength when compared to the first two groups principally associated with residual tensile stress in the surface. Notice that in this case the variation in roughness produced experimentally had no apparent effect on fatigue behavior. In looking at the data for the titanium alloy and for inconel 718, it is seen that under typical chip-making conditions using sharp tooling, variations in surface finish had no effect on fatigue. As indicated before, each of these groups of specimens exhibited virtually identical surface residual stress within the group. An interesting sidelight is shown in the bottom bar in Figure 9 in which inconel 718 was abusively turned using dull tooling. Identification of such areas of low material sensitivity to machining parameters have identified a number of opportunities for realizing significant cost reductions in production, as mentioned earlier in this paper.

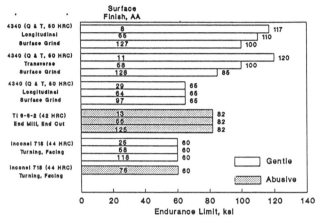

Fig. 9. Effect of surface finish on the fatigue strength of various materials.

A brief comment on the grinding/finishing of bearings is perhaps in order. Much of the information and manufacturing practice on these products is proprietary, hence published data is not readily available. It has been shown, however, that moderate compressive stresses in the surfaces of both ball and roller bearing races have resulted in measurable improvements of rolling fatigue resistance. This leads, of course, to enhanced bearing performance. As a result, residual stress measurement procedures have been incorporated in the production cycle of more than one manufacturer of high performance anti-friction bearings in order to control and assure the fatigue resistance of their products.

Shot peening, because of its wide use in hardware production, is deserving of special consideration in any discussion dealing with the inter-relation of residual stress and fatigue. The literature contains hundreds of references related to peening vs. hardware performance. Springs in particular are a popular component for such discussion. In general terms, peening produces surface zone compression that will be increased as the intensity of the process is increased (harder shot, more exposure time, etc.). Typical residual stress distribution in hardened steel as a result of shot peening is shown in Figure 10. [7] Note that the harder shot (R_c 61) produces the higher level of the compressive residual stress. Note also that both of the residual stress profiles shown here result in stresses somewhat deeper than those associated with surface grinding as shown in Figure 2. Similarly, peening often produces a significant improvement in fatigue strength in materials; a factor of three improvement as a result of peening is shown in Figure 11. [7] In the case of peening, however, the role of residual stress is neither isolated nor straightforward since several factors are simultaneously variable:

(1) Peening does produce surface zone compression as shown in Figure 10.
(2) Peening usually work-hardens the material which also tends to increase its observed fatigue strength.

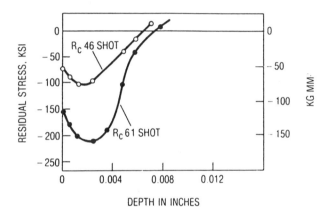

Fig. 10. Peening 1045 steel at HRC 62 with 330 shot.

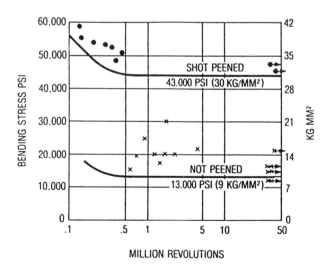

Fig. 11. Fatigue tests on axle shafts.

(3) Peening improves or smooths the surface of rough components such as castings, at least up to a point.

(4) Peening often deteriorates the finish of a machined or ground surface.

(5) Peening may produce laps, microtears and microcracking in a surface.

(6) Peening may cause phase changes in a material, particularly ferrous alloys.

The overall result of peening will depend upon the material and its hardness and the many peening parameters including shot/bead size, hardness, peening intensity, peening angle, exposure time, etc. Generally, this process needs to be evaluated experimentally to determine the optimum process parameters for a given situation. Figure 12 summarizes the effect of extent of peening as measured by Almen intensity on four different materials, CP titanium, Ti-6Al-4V, Al 6061, and Al 7075. [8] Note that in the case of the CP titanium, any amount of peening tends to degrade the fatigue strength in the previously turned and polished surface. Peening of the Ti 6-4 alloy shows substantial fatigue improvement following peening to a moderate level but then subsequent degradation as additional peening cycles are applied. The aluminum alloys also show an initial improvement with peening but then degradation at differing rates with redundant peening cycles. In all of these situations it is probable that the compressive stresses produced by peening are influential in the initial boost to fatigue properties but are then overcome by the detriment of subsequent peening cycles.

In summary, it may be concluded that surface residual stress is frequently a viable predictor of the fatigue behavior of otherwise sound specimens and components. It is equally important to reiterate that the peak stress which is controlling the fatigue behavior may not be located precisely at the surface of the material, hence, a residual stress profile is necessary for a proper evaluation.

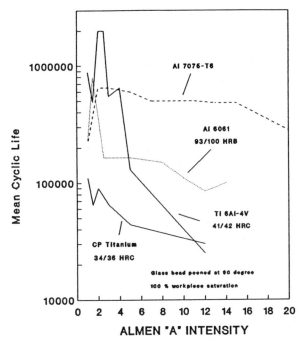

Fig. 12. Fatigue life as a function of shot peen intensity for various materials.

REFERENCES

1. W. P. Koster, et al: *Surface Integrity of Machined Structural Components*, US Air Force Materials Laboratory Technical Report, No. 70-11, March 1970 (ASTIA Document No. AD 870146.)
2. W. P. Koster, et al: *Manufacturing Methods for Surface Integrity of Machined Structural Components*, US Air Force Materials Laboratory Technical Report, No. 71-258, April 1972 (ASTIA Document No. AD 893765L)
3. W. P. Koster: *Surface Integrity of Machined Materials*, US Air Force Report No. 74-60, April 1974 (ASTIA Document No. AD 920055)
4. F. Stablein: *Spannungsmessungen an einsertig abgeloschten Knuppeln*, Kruppsche Monatshefte, Vol. 12, (1931) pp. 93-98.
5. M. E. Hilley, ed: *Residual Stress Measurement by X-Ray Diffraction*, SAE 1784a, Society of Automotive Engineers, New York, New York, 1971.
6. B. D. Cullity: *Elements of X-Ray Diffraction*, Second Edition, Chapter 16, Addison-Wesley, Reading, Massachusetts, 1978.
7. N. K. Burrell: *Controlled Shot Peening of Automotive Components*, Society of Automotive Engineers, Inc., Paper No. 850365, 1985.
8. R. L. Simpson & G. L. Chiasson: *Quantification of the Effects of Various Levels of Several Critical Shot Peen Process Variables on Workpiece Surface Integrity and the Resultant Effect on Workpiece Fatigue Life Behavior*, US Air Force Wright Aeronautical Laboratories Technical Report No. 88-3029, October 1988.

Effects of Residual Stresses on the Fracture Toughness of Zircaloy-2 Tubes

H.P. Mohamadian and A.R. Mirshams
Southern University
Baton Rouge, Louisiana

M.E. Cunningham
Pacific Northwest Laboratory
Richland, Washington

ABSTRACT

Data applicable to evaluating the fracture toughness of Zircaloy-2 pressure tubes were evaluated according to the criteria in ASTM Standards E399 and E813. It was found that the data did not meet the criteria specified in the standards, thus making it necessary to account for residual stress in determining the fracture toughness of the tubes. Therefore, residual stress in pressure tube specimens was experimentally determined to allow the incorporation of residual stresses in the calculation of fracture toughness.

Sections of as fabricated 82.5 mm OD Zircaloy-2 tubes in the 30% cold worked condition were used in testing. Electrochemical Machining (ECM) was used to remove material from the interior of the tube, while measuring the resulting change in strain on the exterior of the tube with delta rosette strain gages. Utilization of this technique to reduce the cross sectional area of a tube without introducing additional stresses offers an accurate procedure for obtaining data to calculate the residual stress distribution.

Assuming elastic anisotropy, the stress equations developed by Voyiadjis, Kiousis, and Hartley were used to determine the residual tangential, radial, and longitudinal stresses as a function of wall thickness. Additionally, these equations allow the determination of the shear stress and subsequently the principal stress profiles. It was found that the measured residual stress level was high enough that it must be accounted for when determining the fracture toughness of the subject Zircaloy-2 tubes.

NOMENCLATURE

ε_z = axial strain
ε_Θ = tangential strain
σ_z = axial stress
σ_Θ = tangential stress
σ_r = radial stress
a = inside radius of tube
b = outside radius of tube
r = instantaneous radius E_r, E_Θ, E_z = Young's moduli of elasticity in r, Θ, and z directions respectively
$\nu_{z\Theta}$ = Poisson's ratio with compression in the Θ direction and tension in the z direction
$G_{z\Theta}$ = shear modulus with the change of angle between z and Θ directions

VARIOUS MANUFACTURING PROCESSES leave residual stresses in structural components due to deformation occurring during the process. The magnitude and direction of the stresses depend on the nature of the deformation process on the materials. Axisymmetric forming operations used to produce rods and tubing typically leave axisymmetric residual stress distributions. In some cases, residual stresses are the predominant factor in crack growth and fatigue causing premature structural failures. Neglecting the superposition of the pre-existing residual stresses and externally applied stresses will produce erroneous crack growth computations.

Zircaloy-2 and Zircaloy-4 alloys currently

are used in the majority of commercial light water reactor fuel rod cladding. These material choices have performed very well with cladding failure rates less than 1% [1]. However, if free flow of water around the fuel rods is restricted due to deformation of the rods, a "hot spot" may develop, leading to potential failure of the rod. Also, the fissioning of the nuclear
fuel produces corrosive gases which, coupled with existing stresses, results in stress-corrosion cracking [2]. Therefore, it is desirable to improve the fuel cladding performance to increase the burnup of the fuel.

An exact method to determine the residual stresses in axisymmetric tubular specimens was proposed by Mesnager in 1919 [3] and further developed by Sachs in 1929 [4]. Voyiadjis, *et al* in 1984 extended the previous analyses of residual stresses to the case where the material exhibits cylindrical elastic anisotropy, i.e. the principal axes of anisotropy correspond to the longitudinal, radial, and circumferential directions of the tube [5]. In this procedure, strains are developed on the outer surface of an axisymmetric specimen in which stressed material is removed from the inner surface. Then the developed strains are measured as a function of radius reduction to be used to calculate the residual stress distribution. Srinivasan et al [6] and later on Rusty and Hartley [7] have shown that electrochemical machining (ECM) provides an excellent method of metal removal without introducing additional internal stresses. However, this technique has its limitations; as the tube thickness become thin ($\approx$ 70% radius reduction), the milling comes to a halt due to pitting of the tube or buckling of the tube wall.

In the following section the results of experiments which employ the ECM process to measure the residual stresses developed on the outer surface of the Zircaloy-2 tubes while material is being removed from the inner surface are described. Since Zircaloy-2 exhibits cylindrical elastic anisotropy, the analyses made by Voyiadjis, et al are used to obtain residual stresses. In the subsequent section, this work will be extended to include the effects of the obtained residual stresses on the fracture toughness data resulting from a fracture toughness testing on the Zircaloy-2 tubes by Pacific Northwest Laboratory (PNL).

RESIDUAL STRESS EVALUATION

A body containing residual stresses but free of external forces and moments, has zero resultant forces and moments on the surface. If successive layers of materials are removed from the body, the residual stress distribution in the remaining material reconfigures to maintain the initial surface conditions. The resulting strains from this reconfiguration are proportional to the stresses in the removed materials. Voyiadjis et al [5], developed an analysis to obtain residual stress distribution in materials fabricated into axisymmetric shapes possessing cylindrical anisotropy.

When internally stressed material is removed, the resulting tangential stress distribution in terms of changing the inner radius from a to r, is expressed as

$$\sigma_\theta(C) = \acute{E}_\theta \left\{ \left(\frac{1-C^{2k}}{2kC^2}\right)\left(\frac{d\theta_o}{dC}\right) - \left(\frac{1+C^{2k}}{2C^{k+1}}\right)\theta_o \right\} \quad (1)$$

Where $C = \dfrac{r}{b}$, $k = \sqrt{\dfrac{E_\theta}{E_r}}$, $\acute{E}_\theta = \dfrac{E_\theta}{1-\nu_{z\theta}\nu_{\theta z}}$

and $\theta_o = \varepsilon_\theta^o + \nu_{z\theta}\varepsilon_z^o$

The radial stress at a radial distance from the tube axis corresponding to the current location of the inner surface is given by

$$\sigma_r(C) = \left\{\frac{\acute{E}_\theta(1-C^{2k})}{2kC^{k+1}}\right\}\theta_o \quad (2)$$

The axial stress distribution is also obtained by relating the resultant of the internal stress distribution to the surface strains yielding

$$\sigma_z(C) = \acute{E}_\theta\left\{\left(\frac{1-C^2}{2C}\right)\left(\frac{d\phi_o}{dC}\right) - \phi_o\right\} \quad (3)$$

Where $\phi_o = \varepsilon_z^o + \nu_{z\theta}\varepsilon_\theta^o$

A residual stress distribution which produces a moment about the tube axis will produce a surface shear strain. The resulting shear stress distribution on the longitudinal-circumferential surface is expressed as

$$\tau_{z\theta}(C) - G_{z\theta}\left\{\left(\frac{1-C^4}{4C^2}\right)\left(\frac{d\gamma_{z\theta}}{dC}\right) - C\gamma_{z\theta}\right\} \quad (4)$$

Where $\gamma_{z\theta}$ - shear strain

EXPERIMENTAL METHOD

MATERIALS - Tube specimens were cut from a section of an as fabricated Zircaloy-2 tube in the 30% cold worked conditions which had a ZrO_2 coating. To establish a good electrolyte conductivity, the coating was removed from the surfaces of the specimens by grit blasting followed by chemical etching. The tubes had an initial outer diameter of 82.5 mm, a wall thickness of 7 mm, and a length of 50 mm. Mechanical properties of the Zircaloy-2 tubes are given in Table 1.

Table 1. Mechanical Properties of Zircaloy-2 at 20°C

Property	Value
Yield Strength	428.0 MPa
Ultimate Strength	713.8 MPa
Flow Stress	571.2 MPa
Young's Modulus	98.25 GPa
Poisson's Ratio	0.355

SYSTEM DESCRIPTION - Figure (1) shows the experimental setup for electrochemical machining. The test section consists of a 316 stainless steel rod (cathode) inserted into a specimen (anode) with the electrolyte flowing through an initial gap of 0.25 mm between the rod and the specimen. The primary variables that affect the material removal were set such that a smooth, uniform milling would be performed. A direct current of 150 mA with a current density of 70 mA/mm^2 was applied across the gap. An acid-proof pump circulated the electrolyte at a flow rate of 3.75 liter per minute (lpm). Suspended solids were collected in a settling tank before the electrolyte was returned to its original reservoir for recirculation. The electrolyte consisted of 25% HCL and 75% 1-Butyl alcohol. To avoid gravitational effects, the electrolyte flowed upwards to insure that the gap separating the rod and the tube was flooded at all times during milling.

PROCEDURE - Since the ECM process obeys Faraday's Law, the mass removal rate was constant. Assuming constant density, the change in the thickness of the specimen was a quadratic function of time. A trial specimen was used to establish thickness reduction as a function of time. The thickness of the specimen was measured by an ultrasonic thickness tester every thirty (30) minutes without interrupting the ECM operation. As shown in Figure 1, four strain-gage rosettes were attached to each specimen at 90-degree intervals and were connected to an OM-941 strain input module for data acquisition. Every thirty (30) minutes, 50 successive readings were taken over an one minute interval. The arithmetic average of each 50 reading batch was stored as a single data point for thirty (30) minutes of milling. A plot of strain as a function of time was constructed for each element of a strain rosette. Finally, strain vs. tube radius reduction was obtained from the strain rate data and material removal rate data by the elimination technique.

RESULTS

Due to a lack of experimental values for the modulus of elasticity in the radial direction, a special case of orthotropy was assumed which occurs in seamless Zircaloy tubing produced by the pilger rolling process. In this material $E_\theta = E_r$, which leads to $v_{\theta z} = v_{z\theta}$. Therefore, equations (1) through (4) were reduced to

$$\sigma_r(C) = \frac{E}{(1-v^2)}\left\{\frac{(1-C^2)}{2C^2}\right\}\theta_o \quad (5)$$

$$\sigma_\theta(C) = \frac{E}{1-v^2}\left\{\left(\frac{1-C^2}{2C}\right)\left(\frac{d\theta_o}{dc}\right) - \left(\frac{1+C^2}{2C^2}\right)\theta_o\right\} \quad (6)$$

$$\sigma_z(C) = \frac{E}{1-v^2}\left\{\left(\frac{1-C^2}{2C}\right)\left(\frac{d\phi_o}{dc}\right) - \phi_o\right\} \quad (7)$$

and

$$\tau_{z\theta}(C) = \frac{E}{2(1+\nu^2)}\left[\left(\frac{1-C^4}{4C^2}\right)\left(\frac{d\gamma_{z\theta}}{dc}\right) - C\gamma_{z\theta}\right] \quad (8)$$

where

$$\begin{aligned}\theta_o &= \varepsilon_\theta + \nu\varepsilon_z = \varepsilon_1 + \varepsilon_2 - \varepsilon_3(1-\nu) \\ \phi_o &= \varepsilon_z + \nu_{z\theta}\varepsilon_\theta = \varepsilon_2 + \nu(\varepsilon_1+\varepsilon_3-\varepsilon_2) \\ \gamma_{z\theta} &= \gamma = \varepsilon_1 - \varepsilon_2\end{aligned} \quad (9)$$

where ε_1, ε_2, and ε_3 are the measured element rosette strains as depicted on Figure 1. Equations (5) - (8) were used to determine stress distributions from the obtained strains at a strain rosette location. The slopes $\frac{d\theta_o}{dC}$, $\frac{d\phi_o}{dC}$, and $\frac{d\gamma_{z\theta}}{dC}$ are calculated at a specified interval of the radius ratio, C, from a computer program using a cubic spline fit to the data. Stress distributions in the radial, axial, and tangential directions for the two test tubes as a function of percentage of tube thickness are shown in Figures (2) - (4). In these figures the thickness is measured from the inner surface of the tube; tensile and compressive stresses are assigned positive and negative values, respectively. The measured strains for each tube are the average reading of each element for the four rosettes mounted 90° apart. It should be noted that the strain readings on different rosettes on a test tube did not vary significantly.

To determine the state of stress for the Zircaloy tube, the principal stresses and the effective stress are given by:

$$\sigma_1, \sigma_2 = \left(\frac{\sigma_z+\sigma_\theta}{2}\right) \pm \sqrt{\left(\frac{\sigma_z-\sigma_\theta}{2}\right)^2 + \tau_{z\theta}^2} \quad (10)$$

and

$$\sigma_{\mathit{eff}} = \frac{\sqrt{2}}{2}\left[(\sigma_\theta-\sigma_r)^2 + (\sigma_\theta-\sigma_z)^2 + (\sigma_r-\sigma_z)^2\right]^{1/2} \quad (11)$$

The principal stresses and the effective stress distributions are shown in figures (6) and (7) respectively.

DISCUSSION

The stress distribution for the two tubes tested have the same pattern although there is substantial difference in stress magnitude in some region of the tube thickness. The radial stress has the smallest magnitude and remains compressive throughout the tube wall and vanishes at the inner surface. The maximum radial stress occurs around 40% of the tube thickness (Fig. 2). The axial stress distribution exhibits compressive stresses from inner surface to approximately 30% of the wall thickness and tensile stress through the remainder of the tube wall reaching a maximum value at about 50% of the tube wall (Fig. 3). The magnitude of the compressive tangential stress is higher than the axial stress but lower for the tensile portion of the stress distribution reaching its maximum tensile stress around 35% of the tube thickness (Fig. 4). The principal stresses are shown in Figures 5 and 6 indicating the maximum compressive stress occurring at the inner surface and maximum tensile stress at about 50% of the tube thickness. The relative high level of effective stress becomes less pronounced with increasing radius, and its maximum lies very close to the inner wall. The maximum value of the effective stress reaches about 30% of the yield strength of the tube in tension (Fig. 7). Although the same testing conditions were applied to the identical tubes, their stress distributions were not similar throughout the wall thickness. This observation is indicative of the non-axi-symmetric behavior of the Zircaloy-2 tubes. Perhaps the most significant trend observed during the run of this experiment was the relatively high compressive stresses observed near the inner wall in the axial and tangential direction.

FRACTURE TOUGHNESS - A set of data, provided by PNL, on fracture toughness of Zircaloy-2 pressure tubes was analyzed by methods prescribed by ASTM-E399 and ASTM-E813. The data did not meet the criteria specified by ASTM-E399 because plane-strain requirements were not satisfied. Therefore, the J_{IC} technique were implemented according to ASTM-E813. Although the estimated value of J_{IC} was valid for two of the specimens, it was invalid for the remaining specimens. An examination of the residual stress levels obtained in this investigation indicate that the magnitudes of these stresses are high enough to affect the results of the fracture toughness data. However, much of the residual stresses are relieved in sectioning the tubes to fabricate the very small fracture toughness specimens (width=12.5 mm, thickness= 6.25 mm). Therefore, the residual stress should be incorporated with the applied stresses in the failure analysis of the tube. Further work is being done to include the effects of residual stress in calculating the applied stress intensity factor, K.

CONCLUSION

The ECM technique had some limitations; as the tube thickness became thin (about 70% radius reduction), the milling came to a halt due to pitting of the tube. By analyzing the stress distributions, the maximum compression stresses were determined to occur at the inner radius of the tube and the tensile stresses reached their maximum values at about 50% of the tube thickness. Therefore, it is evident that within the 70% thickness reduction the maximum tensile and compressive stresses were reached. It was concluded that the measured residual stress levels were high enough that they must be accounted for when determining the applied stress intensity factor of the subject Zircaloy-2 tubes.

ACKNOWLEDGEMENT

The results presented in this paper were obtained in the course of research sponsored by PNL, Richland, Washington, under Subcontract No. 042520-A-A3 to Southern University and its precursors.

The authors appreciate the inputs and suggestions by Dr. E. R. Gilbert and Mr. M. L. Gragg. Also the assistance of Mr. Rory Nettles, Percy Donald, and Reza Paydar is appreciated.

REFERENCES

1. Bailey, W. J., and Wu, S. 1990, Fuel Performance Annual Report for 1988. NUREG/CR-3950, Vol. 6 (PNL-5210), Pacific Northwest Laboratory, Richland, Washington.

2. Hammer, J.J., "A Measurement of Residual Stresses in Zircaloy Tubing," Master Thesis, Louisiana State University. 1982.

3. Mesnager, M., "Method de Determination des Tensions Existant dans une Cylindre Circulaire," Compt. rend., 169, 1391-1393, 1919.

4. Sachs, G., "Der Nachweis Inner Spannungen in Stangen und Rohren," Zeitschriftfur Metalkund 19, 352-357, 1929.

5. Voyiadjis, G.Z., Kiousis, P. D. and Hartley, C. S., "Analysis of Residual Stresses in Cylindrically Anisotropic Materials," SEM, 25(2), 145-147, 1985.

6. Srinivasan, R., Hartley, C.S., and Bandy, R., "Residual Stress Determination in Inconel-600 Tubes Using Electrochemical Machining," Novel Techniques in Metal Deformation Testing, ed. R. Wagoner, TMS-AIME, Warrendale, PA, 163-174, 1983.

7. Rusty J. and Hartley, C. S., "Experimental Measurement of Residual Stresses in Nuclear Fuel Cladding," SEM Conf. New Orleans, LA, June, 1986.

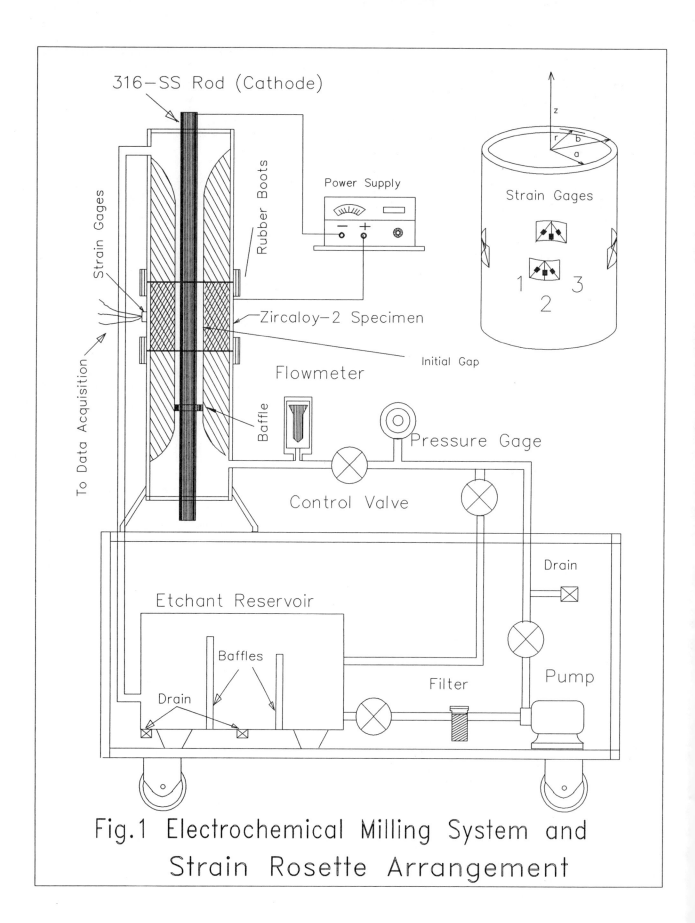

Fig.1 Electrochemical Milling System and Strain Rosette Arrangement

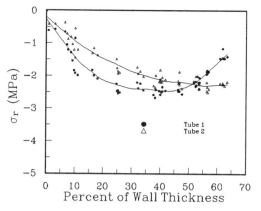

Fig. 2 Radial Stress Distribution for Zircaloy-2 Tubes.

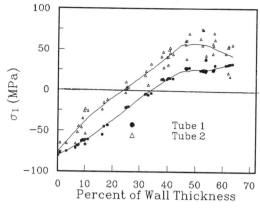

Fig. 5 Max Principal Stress Distribution for Zircaloy-2 Tubes.

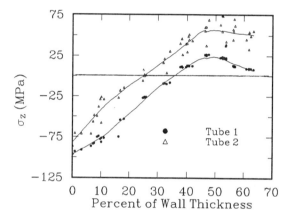

Fig. 3 Axial Stress Distribution for Zircaloy-2 Tubes.

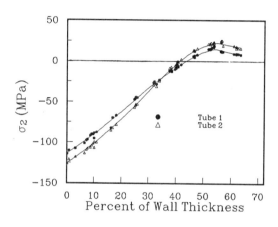

Fig. 6 Min Principal Stress Distribution for Zircaloy-2 Tubes.

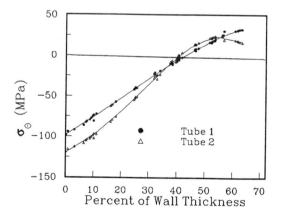

Fig. 4 Tangential Stress Distribution for Zircaloy-2 Tubes.

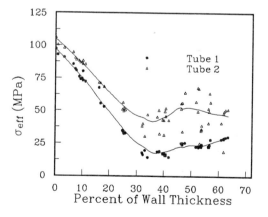

Fig. 7 Effective Stress Distribution for Zircaloy-2 Tubes

Stress and Machinability

J. Tiberg
Swedish Institute of Production Engineering Research
Sandviken, Sweden

THE SWEDISH INSTITUTE
OF PRODUCTION ENGINEERING RESEARCH

ABSTRACTS

Stress in steel affects its machinability. Cold drawn bars have high stress levels in both axial and hoop directions. Stress levels up to 600 MPa tension have have been found. This tensile stress have a beneficial effect on machinability, e.g. on grooving and on parting off operations. Cutting forces are decreased and tool life increases by a factor of 2-4 compared to a bar with low stress or with compressive stress. This is in consequence with what can be expected from the von Mises yield criterium. Straightening in a 2 rollers straightening mill causes the stress in the surface layers to change towards more compressive, i.e. tensile stress change to less tensile or to compressive stress. Therefore straightening decreases the machinability of the bars. The detrimental effect of compressive stress on machinability in an industrial application is reported.
Sachs´method for measurement or residual stress correspons well with Peiter´s method.
The occurence of assymetric stress and their influence on disortion have been known since long. What is new here is the high levels of symmetric stress and its influence of machinability. Therefore specification of stress in steel products will lead to better products, easier to machine and with higher fatigue lives.

RESIDUAL STRESS AND MACHINABILITY - Stress have since long been known to affect the possibility to cut material.

Fig 1. Stress affect cuttability

We have investigated the influence of residual stress in bars on the machinability in grooving operations.

Table I

Bar No	Reduction	Straightening	HB
1	4,5%	Yes	185
3	4,5%	No	185
5	22,5%	Yes	220
7	22,5%	No	220

Chemical composition

C	Si	Mn	P	S
0,17	0,25	1,44	0,020	0,027

TEST MATERIAL

Bars in SAE 1518 were normalized and cold drawn to the same dimension, but the reduction at the drawing operation varied from 4,5% to 22,5%. Some bars were straightened, some were not, see table I. All material were produced from the same cast.

STRESS MEASUREMENT

Two methods for stress measurement were used, the Sachs' boring-out method (1-4) and a new variant of Peiter's bar slitting method (4-5).

THE SACHS' BORING-OUT METHOD - In the boring-out method strain gages are mounded on the bar in axial and hoop directions. Two pairs of cross mounted strain gages were used. By boring and inside turning the bar was reduced from inside in steps. The following symbols will be used.

ε_{ai} = Strain relief in axial direction when boring-out from zero to radius r_i, measured by the strain gage at the surface.

$\varepsilon_{\theta i}$ = Strain relief in tangential direction when boring-out from zero to radius r_i, measured by the strain gage at the surface.

σ_{asi} = Stress relief in axial direction at the surface when boring-out from zero to radius r_i.

$\sigma_{\theta si}$ = Stress relief in tangential direction (hoop stress) at the surface when boring out from zero to radius r_i.

σ_{rsi} = Stress relief in radial direction at the surface when boring out from zero to radius r_i. This is always equal to zero.

σ_{ri} = Stress relief in radial direction at radius r_i when boring out to radius r_i. This is equal to the stress in radial, σr, direction in the solid bar at radius r_i.

σ_a = Residual stress in axial direction in the solid bar.

σ_θ = Residual hoop stress in the solid bar.

σ_r = Residual stress in radial direction in the solid bar.

r = Radius.

D = Diameter of the bar. Here D=38,0 mm.

E = Young's modulus. We have used E=180 GPa.

ν = Poisson's ratio. We have used ν=0.3

The stress relieved at the surface σ_{asi} and $\sigma_{\theta si}$ can be calqulated by Hooke's law.

$$\sigma_{asi} = \frac{E}{1-\nu^2}(\varepsilon_{ai} + \nu\varepsilon_{\theta i}) \qquad (1)$$

$$\sigma_{\theta si} = \frac{E}{1-\nu^2}(\varepsilon_{\theta i} + \nu\varepsilon_{ai}) \qquad (2)$$

From the relieved stress in axial direction, σ_{asi}, the relieved force was calqulated, and by derivation with respect to bored-out area the axial stress σ_a was evaluated through the cross section of the bar.

The radial stress in each layer, σ_{ri}, which is equal to the relieved radial stress can be calqulated by

$$\sigma_{ri} = A_i - \frac{B_i}{r_i^2} \qquad (3)$$

The constants A_i and B_i are found from

$$\sigma_r = A_i - \frac{B_i}{r_i^2} \qquad (4)$$

$$\sigma_\theta = A_i + \frac{B_i}{r_i^2} \qquad (5)$$

Where the relieved stress at the surface when boring-out to radius r_i gives

$$\sigma_{rsi} = 0 = A_i - \frac{B_i}{(D/2)^2} \quad (6)$$

$$\sigma_{\theta si} = A_i + \frac{B_i}{(D/2)^2} \quad (7)$$

Equations (6), (2), (7) and (3) gives σ_{ri}. The hoop stress $\sigma_{\theta i}$ and the radial stress balance according to

$$\sigma_\theta = r \cdot \frac{d\sigma_r}{dr} + \sigma_r \quad (8)$$

THE BAR SLITTING METHOD - Bars were cut longitudinally 25 cm, fig 2, and the bending-out was measured at every 2 cm from the end of the cut. The bending-out was close to circular and the radius of curvature and bending moment, M, was calqulated. This was repeated for another part of the bar after turning down 0,5-2,0 mm. The difference in moment comes from the stress in the turned-away shell and reaction stresses. Consequently the axial stress in the outer zones of the bars were evaluated by stepwise turning and slitting of the bars.

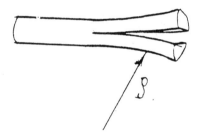

Fig 2 Bending-out of longitudinal cut bars

STRESS ACCORDING TO SACHS - The stress in axial, tangential and radial directions through-out the different bars are given according to the Sachs' method in figurs 3-8.

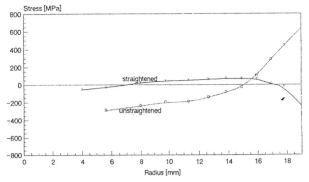

Fig 3 Axial stress in drawn bar No 1 and No 3 reduction 4,5%, straightened and unstraightened.

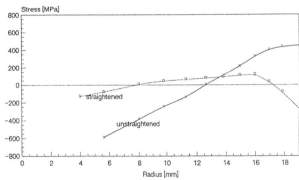

Fig 4 Axial stress in drawn bar No 5 and No 7, reduction 22,5%, straightened and unstraightened.

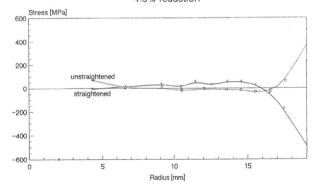

Fig 5 Hoop stress in drawn bar No 1 and No 3, reduction 4,5%, straightened and unstraightened.

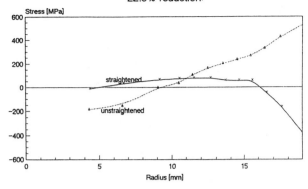

Fig 6 Hoop stress in drawn bar No 5 and No 7, reduction 22,5%, straightened and unstraightened.

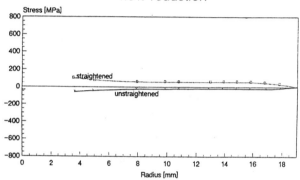

Fig 7 Radial stress in drawn bar No 1 and No 3, reduction 4,5%, straightened and unstraightened.

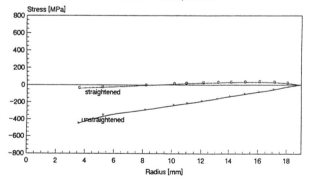

Fig 8 Radial stress in drawn bar No 5 and No 7, reduction 22,5%, straightened and unstraightened.

COMPARISON SACHS' METHOD AND BENDING-OUT METHOD - The results from the boring-out method and the bending-out method were compared in figures 9-10.

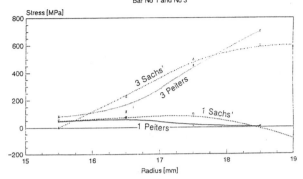

Fig 9 Axial stress in bar No 1 and No 3 according to the boring-out and bending-out methods.

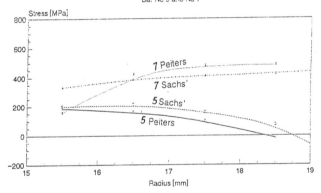

Fig 10 Axial stress in bar No 5 and No 7 according to the boring-out and bending-out methods.

Comments to the stress

- The residual stress measured according the different methods are in good agreement.
- The residual stress were of high magnitude. In the as-drawn condition high tensile stress were found in the surface. Straightening reduced these stress and changed them towards compressive direction.
- The high residual stress in the surface of the as-drawn bars are close to the yield strenght of the material. These stress can be assumed to affect the mecanical behavior of the bar material. Here we have studied the effects of residual stress on machinability.

MACHINABILITY

THEORETICAL CUTTING FORCES - The state of stress in a drawn bar at a grooving operation is illustrated in fig 11 below.

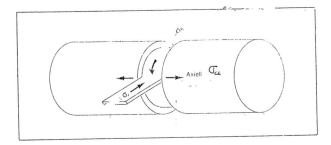

Fig 11 The principal stress in a bar at grooving

In the chip formation zon at the primary shear zon the stress are

σ_a = Axial stress
$-\sigma_\theta$ = Specific cutting force
σ_r = Stress in the chip

These three stresses can be assumed to be principal stresses and we can apply the von Mises yield criterium.

$$Y = \sqrt{\tfrac{1}{2}} \sqrt{(\sigma_a-\sigma_\theta)^2 + (\sigma_\theta-\sigma_r)^2 + (\sigma_r-\sigma_a)^2}$$

Y = Yield stregth of the material.

σ_r, the normal stress in the chip is close to zero. The specific cutting force can therefor easily be evaluated for different axial stress, see fig 12, from where we see that and axial tenseile stress will reduce the specific cutting force. This will probably have a positive effect also on tool life. Compressive axial stress on the other hand can squeeze the tools in grooving and reduce tool life.

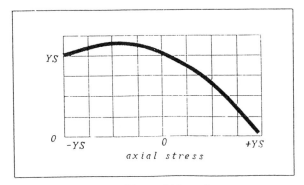

Fig 12 The specific cutting force versus axial stress in grooving.

MACHINABILITY TESTS - The machinability in the different bars presented earlier were tested in grooving with high-speed-steel tools. Groovs, 2,0 mm deep, were cut from the original surface diameter 38,0 mm. Later the bars were turned to diameter 26,0 mm were 3,0 mm deep groves were cut. The chip volyms were the same for both types of grooves, but the stresses were generally lower at 26,0 mm diameter. Cutting data are given in table 2.

Table 2. Cutting data

Diameter	⌀ 38 mm	⌀ 26 mm
Depth of groove	2,0 mm	3,0 mm
Distance between grooves	10 mm	15 mm
Feed, radial	0,05 mm/rev	
Cutting speed	90 m/min	
Cutting fluid	Castrol superedge 10, 5%	
Cutting tool	HSS 6.0 mm width	
Tool life criterium	0,10 mm weare on tool width	

The distance between the grooves is important since the axial stress will fade away near the grooves due to end effects.

RESULTS

TOOL LIFE - The results from the tool life tests are presented in fig 13.

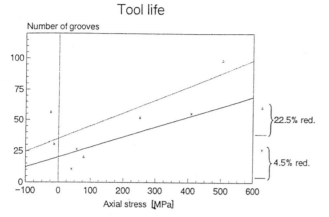

Fig 13 Tool life (number of grooves) versus axial stress is cold drawn bar

CUTTING FORCES - were measuring at grooving:

Cutting speed   50 mm/min
Feed   0,05 mm/rev
Dry cutting   HSS tools 6 mm wide

Four bars were tested near the surface. In one of the bars a test was also made at 4,5 mm below the surface. Due to a lower residual axial stress there, the cutting force increased 40 N, which agrees to the discussion above.

STRESSES AND MACHINABILITY ON SHOP-FLOOR - A factory, producing hollow shafts in automatic screw machines with HSS tools, were used to a tool life of 8 hours. They bought new material from an outside dealer and the tool life was reduced to 2 hours in grooving, external and parting off tools. The shafts were produced from cold drawn leaded resulfurized bar of type SAE 3220 bar. X-ray measurement revield compressive stresses, -175 MPa, in the outer surface of the finished part from the low-tool-life material and tensile stresses, +164 MPa, in the good material (6).

CONCLUSION

The influence of residual stress on machinability has been shown theoretically and experimentally. High residual stress occur in several types of material and they can be positive or negative to machinability. Therefor it is important to find out what residual stress that are present in different bodies to be machined. Moreover, to get optimum machinability the residual stresses must be specified. Residual stress in stainless steel are usually very high, since these steels are generally quenched in water from high temperature to prevent grain boundary corrosion. This high stress is usually compressive in the surface which is often negative to machinability. Residual stress in stainless steel can therefor be one explanation to its bad machinability and to its large variations in machinability. Specification of residual stress in material to be machined is therefor essential.

Table 3. Axial residual stresses and cutting forces at grooving

Bar No	Residual stress	Cutting force
1	- 200 MPa	870 N
3	600 "	770 N
5	- 250 "	800 N
7	450 "	770 N
7 (4,5 mm depth)	350 "	810 N

References

1. Sachs, G, "Zeitschrift für Metallkunde" **19** pp 352-357 (1927)

2. Sachs, G, Espey, G, The Iron Age, sept 18, pp 63-71 (1941)

3. ibid. sept 25, pp 36-42 (1941)

4. Tietz, H-D, Grundlagen der Eigenspannungen, Leipzig (1982)

5. Peiter, A, Werkstattechnik 54 12, pp 597-602 (1964)

6. Tiberg, J, Verkstäderna 16, pp 44-46 (1989)

Effect of Residual Stresses on the Stress-Corrosion Cracking of Austenitic Stainless Steel Pipe Weldments

J.C. Danko
University of Tennessee
Knoxville, Tennessee

ABSTRACT

Boiling water reactor piping systems fabricated from AISI Types 304 and 316 austenitic stainless steels have been susceptible to intergranular stress corrosion cracking. The stress corrosion cracking has been confined to the heat-affected zones of pipe girth welds. Investigations of this cracking phenomenon clearly showed that weld residual stresses were a key contributor to the problem. In addition residual stresses were also introduced in the pipe inside surface by pre and/or post weld machining and grinding operations of the joints. These surface stresses played an important role in the initiation of the stress corrosion cracking. A better understanding of the phenomenon was developed through extensive research and development projects and from these studies methods for mitigating the effects of the residual stresses on stress corrosion cracking were developed, qualified and implemented.

THE PURPOSE OF THIS PAPER IS TO REVIEW THE EFFECT OF RESIDUAL STRESS on the intergranular stress corrosion cracking (IGSCC) of AISI Types 304 and 316 stainless steel (SS) pipe weldments in boiling water reactor (BWR) piping systems.

During the past two decades over a thousand incidents of IGSCC have occurred worldwide in BWR piping systems fabricated from 304 and 316 stainless steels. The cracking has been confined to the weld heat-affected zones (HAZ) in various piping systems (1) (2). Of all the pipe cracking incidents reported, there has been no occurrence of a pipe severance. However, these cracking incidents have resulted in loss of plant availability and revenues, and increased, radiation exposure of craftsmen used in performing inspection and repair operations.

BWR PIPING SYSTEM

The austenitic stainless steel piping systems in BWR models 1 through 6 generally include such lines as the recirculation by pass, core spray, control rod drive return, reactor water clean up and the main recirculation piping. (Figure 1) The piping systems are designed to ANSI Code for Pressure Piping, B 31.1 for the older plants and to ASME Boiler Code, Section III, "Nuclear Power Plant Components" for the newer units. It is important to note, that under BWR operating conditions, and based upon code design stress analyses allowable stresses were presumed not to be exceeded. However, factors such as weld residual stresses are not included in code analyses.

PIPE CRACKING PATTERNS

The IGSCC patterns are quite similar from plant-to-plant. Cracking is confined to the weld HAZs of the pipe girth or butt welds (Figure 2). IGSCC is predominantly of a circumferential orientation; these cracks result from axial stresses. Circumferential cracks usually initiate at a number of sites in the HAZ. These cracks are initially very small and are not connected. In time the length and depth of the intergranular cracks increased. Some of the cracks will link. However based on examination of many plant pipe cracks, leaks or detection by ultrasonic testing (UT) will occur long before 360° circumferential cracking occurs.

On occasion, axial cracks have been observed in the HAZs. Hoop stresses are responsible for the axial intergranular cracks. These cracks, are very small, since they are limited to the HAZ. For this reason axial cracks have been of less concern from a pipe severance consideration. At worst, axial cracks could lead to a leak and be detected. Circumferential cracks on the other hand were viewed during early pipe cracking incidents as having a potential for pipe severance.

No IGSCC has been reported in annealed 304 and 316 SS pipes or in longitudinal seam welds of plate rolled and welded pipes which receive a solution heat treatment (SHT). While there have been a few incidents of transgranular stress corrosion cracking in BWR piping systems, these have been related to off chemistry specifications of the water, namely the chloride content.

RESEARCH AND DEVELOPMENT ACTIVITIES

In response to the worldwide IGSCC incidents, major research and development programs were initiated in 1975 in the USA (3) and Japan (4). BWR vendors, government laboratories and the Electric Power Research Institute participated in broad research programs that included investigations on the mechanism for IGSCC, examinations and analyses of the piping systems, the development and qualification of engineering solutions, methods for non destructive examinations of the pipe welds, leak detection techniques, pipe repair/replacement implementation procedures, and safety concerns. These programs continued through 1987 at a significant support level, and while the effort has tapered off, progress continues on laboratory and field testing of water chemistry engineering solutions.

For purposes of this paper, the research and development results of the effect of residual stresses on the IGSCC are presented along with a model for the mechanism of IGSCC.

MODEL FOR IGSCC

A model for the mechanism of IGSCC in welded Type 304 and 316 SS pipes was developed and subsequent improvements in the model were made (5) (6). The basis for this model was the coexistence of three key factors in the HAZ: (1) a weld sensitized microstructure, (2) tensile stresses or stress intensity, and (3) an environment that supports IGSCC. The IGSCC susceptibility of the welded 304 and 316 SS depends on the severity or intensity of the three contributing factors. A reduction in the severity of one or a combination of these factors or the elimination of any one of these factors alleviates the problem. A schematic representative of these factors is illustrated in Figure 3.

The three key factors may vary from pipe weld-to-pipe weld and from plant-to-plant. These variations result from the chemical composition of the heats of stainless steels used, the thermal/mechanical history in fabricating the pipe, the pipe diameter, the welding procedures welding process, weld joint preparation, pre and post welding operations, the BWR coolant chemistry and temperature. Thus, incidents of IGSCC will vary with time of operation, the pipe diameter, the piping systems and water chemistry.

BWR PIPING STRESSES

In the 1974-1975 period when a rash of IGSCC incidents occurred in 4 in. (100 mm) diameter 304 SS recirculation by pass lines, a comprehensive piping system stress analysis was performed. The results confirmed that the piping systems were within the allowable design stress range. However, weld residual stresses were assumed to experience plastic shakedown during operation and were therefore, not included in the stress calculations. In order to properly address the role of weld residual stresses on IGSCC, the research and development programs at the Electric Power Research Institute and the General Electric Nuclear Energy Division focused on the measurement of residual stresses.

EXPERIMENTAL PROCEDURES

Preparation of Pipe Welds: Small diameter seamless and large diameter plate rolled and welded 304 SS pipes used in the measurements were ASTM A376 and A358, respectively. A number of heats of various pipe sizes in the mill-annealed condition, representative of plant materials were procured for the tests.

Welding parameters used for the 304 SS girth welds were based on typical field welding specifications for the early BWR plants. An extended land type of weld level was used with Grinnel type 308 SS consumable inserts. The internal diameters of all pipe sections were machined counterbored to dimension. Multi-pass welding was performed using gas-tungsten-arc welding (GTAW) of the Grinnel insert; GTAW of the second layer using bare filler wire (ER-308), and shielded-metal-arc welding (SMAW) of subsequent layers using coated electrodes of E308-16. Weld heat input varied with the specific layer and pipe diameter.

Using these procedures the number of weld layers for schedule 80 4 in. (100 mm), 10 in. (254 mm) and 26 in. (660 mm) diameter 304 SS pipes were 7, 14, and 30 passes respectively. Measurements of the temperatures in the welds during the welding process showed variations from pass-to-pass for the different diameter pipes.

Residual Stress Measurements: Residual stress measurements of the austenitic stainless steel pipes used two methods: x-rays and stress relief. X-ray methods were used for the internal surface before and after selective electropolishing or subsurfaces (7). X-rays a nondestructive method, is limited to depths of approximately 5×10^{-4} in (0.02-mm). The accuracy of the x-ray method was ± 5 to 7 ksi (± 35 to 50 MPa)

The stress relief method measures the bulk residual stresses in the region of the surface of the body or through the wall. In this method, strain gages are attached to the outside and inside surface of the pipe weldment and the weldment is carefully sectioned. Based on the amount of stress relief during the sectioning process, the weld residual are calculated.

To measure the throughwall residual stresses, the pipe weldments were cut into thin sections and new strain gages were attached and further sectioning was performed. Calculations of residual stresses from the measured strains were made using isotopic plane-stress, stress-strain relations. During the section of the pipe weldment, a redistribution of the stresses occurs and this was accounted for in the analysis. Accuracy of this method was approximately 15 ksi (135 MPa).

RESIDUAL STRESSES IN AS-WELDED PIPE

Initial measurements on girth welded 304 SS schedule 80 pipe of 4 in. (100 mm), 10 in. (254 mm) and 26 in. (660 mm) diameter showed considerable variations in both the axial and hoop stresses of the internal surfaces. These variations were more prominent in the 4 in. (100 mm) and 10 in. (254 mm) pipe while the fluctuations were less in the 26 in. (660 mm) diameter pipe. Examination of the data, the welding procedures and the properties of the pipe materials did not provide an explanation of these variation. However there was one difference between the pipes, namely the pipe fabrication process. The 4 in. (100 mm) and 10 in. (254 mm) pipes were seamless and produced by an extrusion process while the 26 in. (660 mm) pipe was plate rolled, welded and solution heat treated. A clear explanation for the observed variations was not developed.

Since the initiation of IGSCC occurs in the weld HAZ of the pipe inside surface, axial residual stresses as a function of distance from the weld centerline were measured by the combination of x-ray and strain gage measurements. The results of the measurements on the three pipes sizes are plotted in Figure 4. As shown in this figure high tensile axial stresses existed on the internal pipe surfaces in the HAZ with the highest peak axial stresses in the 4 in. (100 mm), intermediate in the 10 in. (254 mm) and the lowest in the 26 in. (660 mm) diameter pipe. These weld residual stress also showed a rather steep gradient as a function of distance from the weld centerline and were generally axisymmetrical. At approximately 0.3 in. (7 mm) from the weld centerline, the peak axial stresses changed from tensile to compressive. The compressive residual stresses reached a level of approximately -70 ksi (-500 MPa) at 0.4 in. (10 mm) from the weld center line for three different pipe sizes.

In the examination of pipe welds removed from operating BWRs, IGSCC often followed the machining grooves in the counterbore. Also some of the weld counterbores were ground and cracking propensity appeared to be more pronounced. It was determined that grinding operations were commonly used for preparing weld counterbores and also for post-welding operations. As shown in Figure 4, very high tensile axial surface stresses were measured in the HAZs of the machined counterbore surfaces. Because of the number of IGSCC incidents in heavily machined and ground counterbores, additional x-ray residual stress measurements were made on machined and ground surfaces of 304 SS.

Machining and in particular grinding of the pipe surfaces produces a cold worked surface layer along with surface defects that may be potential sources of IGSCC initiation. Cold working also results in higher yield stress accompanied with lower strains to fracture. The results of x-ray residual stress measurements on unwelded pipes and flat coupons are reported in Table I. for 304 SS.

The results presented in Table I show the importance of surface treatment on the surface residual stresses. Residual stresses are higher in the direction parallel to the lay than in the perpendicular orientation. The high surface tensile residual stresses contribute to the initiation of IGSCC.

Samples of the 304 SS coupons with ground only and heavily machined plus ground surfaces were used to obtain residual stress profiles. High tensile stresses parallel to the lay were produced on the surface. Removal of the surface by electropolishing was used to obtain subsurface measurements. While the tensile stresses in the hand ground surface diminished rapidly to a few ksi (14 MPa) at a depth of 0.006 in. (0.150 mm), the machined and ground surface tensile residual stresses of approximately 30 ksi (210 MPa) persisted to depths greater than 0.014 in. (0.35 mm). These results show the importance of the type of surface preparation procedures on the state of residual stress.

Since the BWR piping systems operate at 550°F (288°C) during full power operation, it was important to determine the effect of temperature on the thermal relaxation of residual stresses. Surface x-ray residual stress measurements were made at room temperature on abusively ground samples before and after exposure to heat treatments at 571°F (300°C), and 932°F (599°C) for various times in vacuum. The stress direction was parallel to the grinding direction. At the normal BWR operating temperature of 550°F (288°C), there was essentially no change in the tensile residual stresses after 350 hours. This indicates permanence of the weld residual stresses in the 304 SS pipe weldment. However, at 932°F (500°C) which is well above the BWR operating temperature there was a significant reduction in the tensile residual stresses showing thermal relaxation of the surface residual stresses.

The throughwall distribution of weld residual stresses of three different pipe butt welds of 4 in. (100 mm) schedule 80 304 SS pipes were examined. These pipes all showed the same azimuthal variation in axial and hoop stress on the internal surface. A similar weld residual stress pattern was observed for the throughwall distribution measurements. The residual stresses were predominantly tensile through the center wall near the weld fusion line, a

TABLE I

Effects of Surface Treatment on Surface Residual Stress in 4 in. (100 mm) 304 Stainless Steel Pipe

Inside Surface Preparation	Maximum Surface Residual Stress ksi (MPa)	
	Circumferential	Axial
Machine	80 (550) T	10 (70) T
Light Grind (Hand Grinder)	80 (550) T	0 (0)
Heavy Grind (Hand Grinder)	80 (550) T	17 (120) T
Shot Peened	42(290)C	
As Received	15(150)C	

T = tension C = Compression   Grinding in circumferential direction

region where sensitization would be expected to be most severe.

For the 26 in. (660 mm) diameter 304 SS pipe, the axial weld residual stresses varied from moderately tensile on the internal surface to compression at about ten percent of the wall thickness. Peak compressive stresses where produced at approximately 25% of the wall thickness and the stresses remained compressive to about 75% of the wall thickness before going tensile again. Throughwall weld residual stress measurements for the 26 in. (660 mm) pipe at a distance of 0.13 in. (3.3 mm) from the weld centerline is illustrated in Figure 5. Subsequently, other laboratories performed measurements of the throughwall weld residual stresses in large diameter pipes and the results were in good agreement. These results showed significant differences in the throughwall residual stresses relative to the 4 in. (100 mm) diameter 304 SS pipes. A difference that appears to be manifested in the IGSCC propensity in BWR piping.

In order to determine the permanence of the as-welded residual stress, measurements of the residual stresses were performed on a 10 in. (254 mm) and 24 in. (610 mm) diameter 304 SS pipe removed from plant service (8). The results showed that peak internal surface axial stresses on the 10 in. (254 mm) pipe were 60 ksi (430 MPa) which compared favorably with the values of 50 ksi (345 MPa) for an as-welded pipe of the same diameter. Throughwall residual stress distribution of the 24 in. (610 mm) pipe removed from service after 10 years of operation because of IGSCC is included in the data of Figure 6. The data fit extremely well in the plot of the as-welded pipes of 304 SS. These results clearly demonstrate that the weld residual stresses are permanent and do not shakedown during the BWR operations at 550°F (288°C). This is consistent with the IGSCC incidents in BWR piping systems after years of operation.

ALTERING WELD RESIDUAL STRESSES TO MITIGATE IGSCC

Once the model for the phenomenon of IGSCC was developed, a number of potential engineering solutions for the mitigation of IGSCC were identified. Some of the measures were based on altering the well residual stress distribution. A brief description of these measures presented in chronological order of their development follows.

Heat Sink Welding: One of the first measures developed was heat sink welding (HSW). This was intended to reduce sensitization, but the benefits for improved resistance to IGSCC was determined to be related to altering the residual stresses in the pipe weld. In HSW the root pass and the first few weld pass layers are made to provide structural support of the pipe weld. This operation is performed by manual gas tungsten arc welding (GTAW). Subsequent weld layer deposits are made while flowing or sprayed water is applied on the inside of the pipe. ≥This process resulted in compressive residual stresses on the pipe inside surface and partially throughwall. The peak surface axial residual stresses of an as-welded and heat sink welded pipe of 10 in.(254 mm) 304 SS is shown in Figure 7. Alteration of the residual stresses in the pipe butt weld is clearly illustrated. While HSW did in fact reduce the weld sensitization, the critical level to mitigate IGSCC was not achieved. (9) (10) (11)

Induction Heating Stress Improvement: Induction heating stress improvement, (IHSI) a process that uses induction heating of the pipe weldment was conceived and developed in Japan (12). In this process the outside surface of the pipe weldment is heated to 930 to 1020°F (500 to 550°C) while the pipe inside surface is maintained at 212°F (100°C) with flowing water. The resultant temperature gradient between the outside and inside surface of the pipe produces a small thermal strain and the as-welded residual stresses are altered. The final state of stresses are compressive on the inside surface and partially throughwall on the inside of the pipe. Typical compressive residual stresses imparted by IHSI for a 16 in. (406 mm) diameter 304 SS pipe weld are illustrated in Figure 8 (13). An analysis of the relaxation of the residual stress produced by IHSI showed some relaxation with prolonged time at the BWR operating temperature (550°F-288°C), but the compressive residual stresses remained. IHSI of pipe welds with IGSCC, did not result in crack extension. Moreover, stress analysis of this case did not show much of a change in the stresses of the tip of IGSCC with the compressive residual stress zone. IHSI was first implemented in the Japanese BWR piping systems. Subsequently, this process was utilized in American and European BWRs.

Weld Overlay: Weld overlay (WO), a process developed to repair pipes with IGSCC provided structural reinforcement of the pipe weldment to meet the original code stresses. A schematic of WO on a pipe with IGSCC is shown in Figure 9. The WO also introduced compressive residual stresses on the inside diameter of the pipe. This is a result of the shrinkage of the pipe from the solidification of the weld overlay deposit. The favorable residual stresses contributed to the reduction in crack growth or arresting existing IGSCC cracks. However, WO may not always produce compressive residual stresses. In WO of 28 in. (716 mm) diameter 304 SS pipe, tensile and compressive residual stresses were measured. This information indicates that WO will not always result in compressive residual stresses, particularly in large diameter thickwalled pipes.

In the WO process, a 308 L filler material with a starting ferrite number (FN) of 11-17 is selected in order to produce an as-deposited FN of 8. Low heat inputs are used for the initial deposit to minimize dilution while higher heat inputs are used for subsequent weld layer deposits. The reason for a minimum delta ferrite level of FN8 is to arrest IGSCC in the WO. In the application of WO, remote automatic GTAW is used and in many cases the reactor water coolant remains in the pipe. Under these conditions, the reactor coolant is not drained from the piping system; this results in significant economic benefits.

Last Pass Heat Sink Welding: Last pass heat sink welding (LPHSW) is a modification of the HSW; the cooling water is not introduced until the last pass or crown weld deposit is applied. During the last pass, a very high heat input is used to establish a large temperature gradient across the pipe wall. This is analogous to the gradients achieved in the IHSI process and results in similar residual stress distributions. A plot of throughwall residual stresses for a 24 in. (610 mm) diameter 304 SS processed by LPHSW is illustrated in Figure 10. A comparison with Figure 10 for the IHSI treated pipe shows the similarities in the residual stress patterns. (14)

Mechanical Stress Improvement Process: The most recent method developed for altering the as-welded residual stress distribution is the mechanical stress improvement process (MSIP). This is accomplished by a special tool consisting of a mechanical clamp connected with circumferential studs which are tightened by a hydraulic tensioner. (15)

As the tensioners are hydraulically pressurized, the halves of the clamp load the pipe circumference elastically and then plastically. The amount of deformation is precisely controlled. The operation is very simple, is accomplished in 15 to 20 minutes and can be repeated if necessary. In applying the MSIP, the pipes may be filled with reactor coolant or emptied. Deformation of the pipe extends throughout the weldment, and the reduction in the pipe diameter imparts residual compressive stresses at the inner pipe surface in both axial and hoop directions. The residual compressive hoop stresses extend through the entire pipe wall. MSIP does not introduce tensile stresses on the inner surface of the pipe and if any IGSCC is present it will not propagate. Residual stress measurements confirm very favorable compressive residual stresses in MSIP processed pipe weldments.

PIPE TESTING OF RESIDUAL STRESS MITIGATION MEASURES

In order to determine of effects of residual stresses on the IGSCC of 304 SS and 316 SS pipe weldments, testing of full size welded pipes was necessary. Small laboratory test specimens could not be used since stress relaxation occurs when the specimens are removed from the pipe weldments. To test the pipes special pipe test stands were designed and fabricated.(16) These test units located in a number of different laboratories were capable of testing pipes ranging from 4 in. (100 mm) to 24 in. (610 mm) in diameter.

A statistical test program was planned using a number of different heats of 304 SS pipes in the as-welded and in the as-welded with the various residual stress mitigation methods applied. Thus a comparison of the times-to-failure could be made and a factor of improvement (FOI) over the as-welded reference pipe welds could be calculated. (17)

The pipe test specimens were tested under accelerated conditions to shorten the times to failure. These conditions included high stress levels above the base material yield stress at 550°F (288°C), high dissolved oxygen contents of 8 ppm in the high purity water and cyclic loading of 0.75 cycles per hour using a trapezoidal wave form.

All of the residual stress mitigation measures with the exception of MSIP and WO were tested. Some of the test were of limited duration and the results showed FOI values for HSW, IHSI and LPHSW of 15, >10 and >4, respectively (Ref. 21). These results show the benefits of the residual stress mitigation measures.

DISCUSSION

The model for IGSCC of welded pipes of 304 and 316 SS identified three key contributors: a weld sensitized microstructure, the BWR water coolant and tensile stresses. During the early time period of IGSCC of the BWR piping systems, the common piping materials were 304 and 316 SS. For the common grades the carbon content was greater than 0.035 W/O. On the average, the carbon levels were 0.05 W/O. At this level, sensitization is to be expected in the weld HAZ. New nuclear grade material (18), (304 and 316 NG) with carbon levels of 0.02 W/O maximum do not sensitize during welding, and therefore, are extremely resistant to IGSCC.

The BWR water coolant has small amounts of dissolved oxygen (200-300 ppb) at the BWR operating temperature of 550°F (288°C). At these oxygen levels, IGSCC will occur given the presence of the other two contributors to IGSCC. Current research is focused on hydrogen water chemistry which consists of hydrogen addition to the feedwater to reduce the dissolved oxygen contents to 20 ppb and less. At these oxygen levels IGSCC may be avoided. (19)

Finally, tensile stresses above the yield stress of the base material are required to have the necessary conditions for IGSCC of weld HAZ. Piping systems designed to code stresses ordinarily would not be prone to IGSCC. However, the code stress analysis do not include the weld residual stresses. When the weld residuals are considered in the total stresses of the pipe welds, the conditions are present for IGSCC. The stresses are exacerbated by the tensile residual stresses that result from machining and grinding operations that may be performed on the pipe welds. These operations also cause surface damage such as tearing, grooves, laps and seams. In addition, the weld residual stresses and the surface residual stresses resulting from surface preparation operations show little relaxation at the BWR operating temperature of 550°F (288°C). Thus these residual stresses become a permanent part of the total applied stress acting of the pipe welds.

IGSCC occurs in the weld HAZ. The reasons for this are the presence of a sensitized microstructure, and the existence of high tensile weld residual stresses and the BWR coolant. Although the weld residual stresses are greater in the weld metal, the duplex microstructure consisting of austenite and small amounts of delta ferrite is extremely resistant to IGSCC. When sensitization does occur in this microstructure, the precipitation of chromium carbides occur at austenite-ferrite boundaries and not at austenite-austenite grain boundaries that is typical of the HAZ. NUREG 1.44 specifies a minimum of 5% delta ferrite in the weld metal for better weldability. At these ferrite levels, the weld metal has considerable resistant to IGSCC. Thus, this guide also provides a benefit for mitigating IGSCC. While there have been some cases of IGSC cracks in weld metal, these have occurred in regions with low delta ferrite contents on order of a few percent and less. However, the larger IGSC cracks and leaks in pipe weldments occur in the HAZ, the path of minimum resistance.

The distribution of weld residual stresses in the various pipe sizes are likely responsible for the frequency of IGSCC incidents. For example, IGSCC incidents were more prominent in small diameter pipes (4 in. (100 mm)) In time the IGSCC events appeared in 10 in. (254 mm) and 16 in. (510 mm) pipes. Finally after years of operation, IGSCC was detected in large diameter pipes >20 in. (510 mm). It is important to note, that there has been no reported pipe severance in over a 1000 IGSCC incidents worldwide. This is related in part to the non uniformity in the tensile stresses acting on the pipe weldment and variations in the level of sensitization in HAZ.

Once the contribution of weld residual stresses to IGSCC of pipe weldments was recognized, methods for the mitigation of IGSCC by altering the residual stresses were developed, qualified and implemented in pipe repair/replacements. These mitigation measures included: HSW, IHSI, WO, LPHSW and MSIP. All of these measures produce compressive residual stresses on the inside pipe surface and partially throughwall. These residual stress measures in pipe welds without IGSCC were very effective in mitigating IGSCC. The measures also were effective in pipe welds with IGSCC by arresting crack propagation or significantly reducing crack growth. The application of a given mitigation, process depends on many factors that may include: the number of pipes with IGSCC, a leaking vs non-leaking pipe weld, the diameter of the pipe, location in the piping system, scheduled vs. non-scheduled outage, radiation levels at the IGSCC site, availability of skilled craft, ease of repair, time estimate for repair/replacement and review and approval of the Nuclear Regulatory Commission (NRC).

All of the mitigation measures have been implemented for repair/replacement of 304 and 316 SS pipes with IGSCC. Of the various process, HSW and LPHSW have found limited application, while WO and IHSI have been used in many plants. MSIP, a relatively new measure has been used in some plants and may have broader applications in the future. In considering the use of the mitigation methods, the position of the NRC as presented in NUREG 0313 Revision 2 (20) regarding the acceptance and inspection requirements is a guiding factor. For example, the frequency of pipe inspections and acceptance by the NRC of temporary or long term fixes are of paramount importance to the utilities. This of course, will have an impact on plant outages, radiation exposure of skilled craft performing the inspection and repair/replacement of piping.

SUMMARY

IGSCC incidents in the weld HAZs of AISI Types 304 and 316 SS piping systems in BWRs have had a negative impact on plant availability and economics worldwide. In early investigations of the problem, a model for the mechanism of IGSCC was developed. Tensile stresses were identified as one of the three major contributors. However, weld residual stresses were not included in the code stress analyses of the piping systems. Measurements of the weld residual stresses in 304 SS pipes revealed high axial tensile stresses on the pipe inside wall and partially throughwall. In addition, tensile residual stresses were created by machining and grinding operations used in the preparation of the welds. The weld residual stresses and those resulting from surface preparation were permanent and did not shakedown during plant operation. These residual stresses when combined with the operational stresses exceeded the yield stress of the base material and contributed in a very major way to the IGSCC of the pipe welds.

With this clear understanding of the effects of residual stresses on IGSCC,. a number of mitigation measures were developed which altered the weld residual stresses. These measures or processes included; HSW, IHSI, WO, LPHSW and MSIP. After the application of these processes, the tensile residual stresses on the inside pipe surface and partially throughwall are reduced significantly or become compressive. Under these stress conditions initiation of IGSCC is extremely difficult and for existing IGSC cracks, propagation is arrested or substantially reduced.

In order to qualify the benefit of the mitigation measures, full size welded pipes were tested under simulated BWR conditions. The times-to-failure of the pipe welds using the various mitigation methods were compared to the as-welded pipes of the same heats of 304 and 316 SS. HSW, IHSI and LPHSW processed pipes showed significant improvement in times to failure. Pipe test results were not available for WO and MSIP.

Of the various stress mitigation methods implemented, the most widely used for pipe repair/replacement were WO and IHSI. MSIP a more recently developed process is finding increased application while HSW and LPHSW have had limited use.

ACKNOWLEDGEMENT

The results reported in this manuscript were to a large extent generated under the joint Electric Power Research Institute - Boiling Water Reactor Owners Group Research Program on Pipe Cracking. Other sources of information were General Electric Company, Hitachi, Ltd. Toshiba Corporation and Ishikawajima-Harima Heavy Industries Co., Ltd.

REFERENCES

1. Technical Report, Investigation and Evaluation of Cracking in Austenitic Stainless Steel Piping of Boiling Water Reactors, NUREG-75-1067, Oct. 1975, Nuclear Regulatory Commission.

2. H. H. Klepfer, et al, "Investigation of Cause of Cracking in Austenitic Stainless Steel Piping, NEDO-2100, Vol 1, General Electric Co., July 1975.

3. R. E. Smith, J. C. Danko and A. D. Rossin, BWR Intergranular Stress Corrosion Cracking Research Program Overview, American Power Conf. Paper No. 4, April 1980.

4. Y. Ando, Overview of BWR Pipe Cracking in Japan, EPRI WS-79-174, Vol. 1 May 1980.

5. R. E. Hanneman, R. E. Rao and J. C. Danko, Intergranular Stress Corrosion Cracking Cracking in 304 SS BWR Pipe Welds in High Temperature Aqueous Environments, TMS-AIME, 1977.

6. R. E. Hanneman, Models for the Intergranular Stress Corrosion Cracking of Welded Type 304 Stainless Steel Piping in BWR, EPRI WS-79-174, Vol. May 1980.

7. Studies on AISI Type 304 Stainless Steel Piping Weldments for use in BWR Application, EPRI NP-994 Dec 1979.

8. Measurements of Residual Stresses in Type 304 Stainless Steel Piping Butt Weldments, NP-1413, EPRI 449-1, June 1980.

9. R. M. Chrenko, Residual Stress Measurements on Type 304 Stainless Steel Welded Pipes, EPRI WS-79-174, Vol. 1, May 1980.

10. Controlling Residual Stresses by Heat Sink Welding, EPRI NP-2159-LD, Dec. 1981.

11. R. Sasaki et al, Mitigation of Inside Surface Residual Stresses of Type 304 Stainless Steel Pipe Welds by Inside Water Cooling Method, EPRI WS-79-174, Vol. 1, May 1980.

12. S. Tanaka and T. Umemoto, Residual Stress Improvement by Means of Induction Heating, EPRI WS-79-174, Vol. 1, May 1980.

13. Induction Heating Stress Improvement Qualification, EPRI WS-79-174 Vol. 1, May 1990.

14. Experimental Determination of the Effect of Last Pass Heat Sink Welding on Residual Stress in a Large Diameter Stainless Steel Pipe, EPRI NP-3361 Nov. 1983.

15. J. S. Porowski et al Mechanical Stress Improvement Process for Altering Weld Residual Stress, Nuclear Eng., April 1985.

16. J. C. Danko, et al, A Pipe Test Method for Evaluating the Stress Corrosion Cracking Behavior of Welded Type 304 Stainless Steel Pipe, ASME Dec. 1978.

17. R. Post and J. C. Lemaire, Statistical Approach to Qualify Countermeasures, EPRI WS-79-174, Vol. 1, May 1980.

18. J. Alexander et al, Alternative Alloys for BWR Pipe Applications, EPRI NP-36 71-LD, Oct. 1982.

19. BWR Hydrogen Water Chemistry Guidelines: 1987 Revision, EPRI JNP-4947 SR, Dec. 1988.

20. U.S. Nuclear Regulatory Commission, Technical Report on Material Selection and Processing Guidelines for BWR Coolant Pressure Boundary Piping. NUREG-0313 Rev. 2.

21. R. L Jones, Summary of Stress and Sensitization Remedies, EPRI NP-3684 SR Vol. 2, Proc. 2nd. Seminar on Countermeasures for Pipe Cracking in BWRS, Sept. 1984.

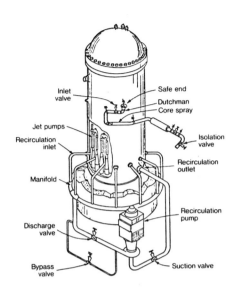

1. Schematic of a BWR recirculation piping system.

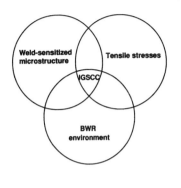

3. Major factors contributing to IGSCC in welded 304 and 316 SS pipes. (Ref. 5).

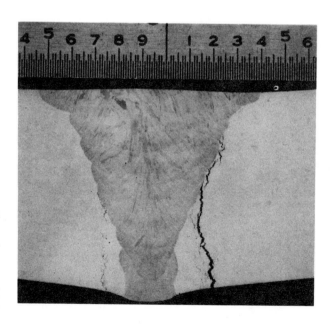

2. Macrograph of IGSCC in weld HAZ of pipe removed from operating BWR.

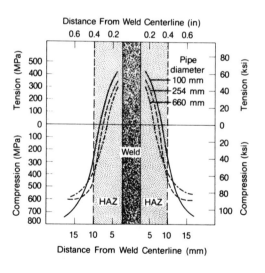

4. Peak axial surface tensile residual stresses in welded 304 SS pipes of 4 in. (100 mm), 10 in. (254 mm) and 26 in. (660 mm) diameter. (Ref. 7).

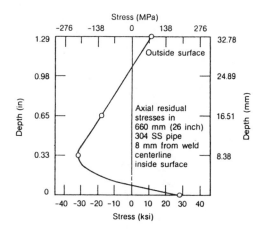

5. Throughwall axial residual stresses for welded 304 SS girth weld in a 26 in. pipe (660 mm) diameter (Ref. 8).

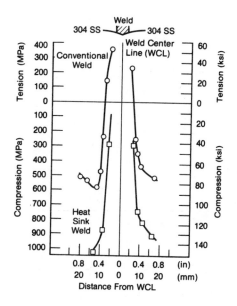

7. Total axial internal surface residual stresses for 304 SS 10 in. (254 mm) pipe welded conventionally and by HSW. (Ref. 9).

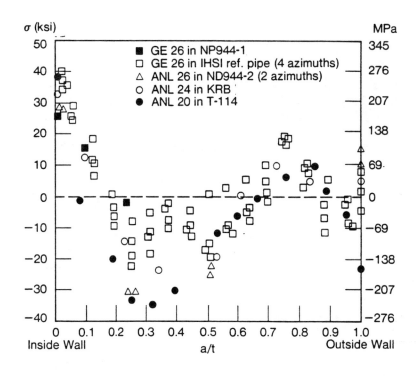

6. Throughwall residual stresses in large diameter 304 SS pipe welds. (Ref. 8).

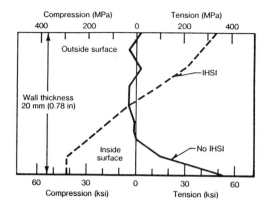

8. Residual stresses in 16 in. (406 mm) 304 SS pipe weld with and without IHSI. (Ref. 13)

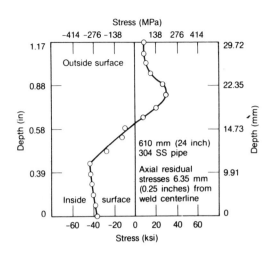

10. Weld residual stresses in 24 in. (610 mm) 304 SS pipe weld produced by LPHSW. (Ref. 14).

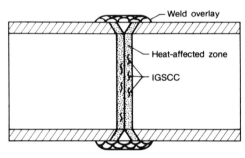

9. Schematic of WO on pipe with IGSCC.

Assessment of Component Condition from X-Ray Diffraction Data Employing the Sin²-Psi Stress Measurement Technique

E.B.S. Pardue and L.A. Lowery
Technology for Energy Corporation
Knoxville, Tennessee

ABSTRACT

Measuring residual stresses in components is an important tool for process control, quality control, design assessment, and failure analysis. The x-ray diffraction technique is generally a non-destructive technique that provides absolute values for surface residual stresses. It can be an even more powerful diagnostic tool because of the additional information it provides on component metallurgical condition.

Information obtained using the $\sin^2\psi$ technique can enhance the impact of the residual stress value by providing intensity variation, diffraction peak breadth, and d-spacing versus $\sin^2\psi$ data. These data can be directly related to grain size, preferred orientation, surface cold working or hardness, and process changes among other factors. The technique is also powerful in that geometric effects and sample misalignment are readily seen from the numerical and graphical data.

This paper addresses the additional analysis information available from the $\sin^2\psi$ technique. Supporting data are presented from a variety of components.

SURFACE RESIDUAL STRESS can be measured by several x-ray diffraction techniques, i.e., the single-exposure, the double-exposure, or the $\sin^2\psi$ technique. The $\sin^2\psi$, or multiple-tilt, technique not only measures residual stress but also can provide valuable insight into the metallurgical condition of a component.

In the current work a portable instrument employing a position-sensitive detector was used. Some of the examples and the analysis of component condition are applicable, therefore, only to instrumentation of similar configuration and diffractometer radius.

THEORY

The basic equation for stress determination from diffraction data is:

$$\frac{d_{\phi\psi} - d_0}{d_0} = \frac{1+\nu}{E} \{\sigma_{11}\cos^2\phi + \sigma_{12}\sin2\phi + \sigma_{22}\sin^2\phi - \sigma_{33}\} \sin^2\psi$$

$$+ \frac{1+\nu}{E}\sigma_{33} - \frac{\nu}{E}(\sigma_{11} + \sigma_{22} + \sigma_{33}) \quad (1)$$

$$+ \frac{1+\nu}{E} \{\sigma_{13}\cos\phi + \sigma_{23}\sin\phi\} \sin2\psi$$

where

$d_{\phi\psi}$ = d-spacing at ψ
d_0 = unstressed d-spacing (approximated by the d-spacing at $\psi = 0$)
ν = Poisson's ratio
E = x-ray elastic modulus
σ_{xy} = stress component in an orthogonal coordinate system
ϕ = direction in which stress is being measured

If stresses in a component are assumed to be biaxial, then the above equation can be simplified to:

$$\frac{d_{\phi\psi} - d_0}{d_0} = \frac{1+\nu}{E}\sigma_\phi\sin^2\psi - \frac{\nu}{E}(\sigma_{11} + \sigma_{22}) \quad (2)$$

This equation gives a linear response to data plotted as d-spacing versus $\sin^2\psi$. In cases where the data are not linear, the underlying cause of the nonlinearity can be used to assess component condition.[1]

The shift in the diffraction peak is used to measure residual stress whereas the width of the diffraction peak can be used as a measure of plastic deformation. Therefore, the peak width can also be used in assessing component condition.[2]

NONLINEAR d-SPACING VERSUS $\sin^2\psi$ PLOTS

USING POSITIVE ψ ANGLES ONLY - One of the most common types of nonlinear d-spacing versus $\sin^2\psi$ plots encountered is shown in Figure 1. This undulating curve could indicate a large grain size, preferred orientation, or a highly disturbed or roughened surface. If the integrated intensity of the diffraction peak at each ψ angle is uniform, then this probably indicates that the component surface is highly disturbed or roughened. On the other hand, if the diffraction peak integrated intensity varies with angle, then one can assume large grain size and/or preferred orientation conditions exist. It is possible to distinguish between these two conditions by making an additional measurement. Simply translate the component a short distance, say ½ inch, and repeat the measurement in the same direction as the previous measurement. Large grain size is indicated if the integrated intensity versus $\sin^2\psi$ does not follow the same pattern for the two measurements. If the integrated intensity versus $\sin^2\psi$ does follow the same pattern, then the component has preferred orientation. Examples of these conditions are shown in Figures 2 and 3. In cases where large grain size and preferred orientation are indicated by the oscillatory nature of the d-spacing versus $\sin^2\psi$ plots, improvement in the data precision can be obtained by using ψ oscillation with systems using position-sensitive detectors.[3]

Integrated intensity versus $\sin^2\psi$ profiles have been used to identify forming processes and sort components. For example, an extruded aluminum alloy component was identifiable by its unique processing texture (preferred orientation) when compared to the same component formed by another process. Thus the preferred orientation signature was used to sort extruded versus non-extruded parts.

In magnetron applications it is critical to have the components made from the same steel sheet and oriented the same with respect to the rolling direction. Again integrated intensity versus $\sin^2\psi$ data were used to sort components.

Another type of nonlinear d-spacing versus $\sin^2\psi$ plot consists of curvature near $\psi = 0$ (Figure 4). This type of plot indicates a stress gradient perpendicular to the surface of the component or the existence of shear stresses. (Since shear stresses will be discussed in the next section, only stress gradients will be covered here.) This nonlinearity is generally small and may not affect measurement precision significantly. The curvature can be minimized if a radiation with shallower penetration can be used. For example, in aluminum and nickel alloys, chromium radiation will be less sensitive to stress gradients than copper radiation because of shallower penetration by the softer (chromium) radiation.

A third type of nonlinear d-spacing versus $\sin^2\psi$ plot may not exhibit any set pattern. This response has been seen on low stress or stress free components that have a convolution and deconvolution of the $K_{\alpha 1}$ and $K_{\alpha 2}$ components of the diffraction peak at different ψ angles. With some position-sensitive detector systems, this nonlinearity can be minimized by using ψ-angle oscillation (Figure 5). The example shown is a stress-free powder with preferred orientation. In this case, the oscillation by minimizing the orientation effects has resulted in deconvoluting the $K_{\alpha 1}$ and $K_{\alpha 2}$ components. Curve fitting routines that account for the $K_{\alpha 1}$ and $K_{\alpha 2}$ components, of course, will minimize and possibly remove this nonlinearity.

Another type of nonlinearity relates to the geometry of the component instead of its metallurgical condition. It is included here as an aid in data interpretation. Samples with complex geometries, such as gears, often shield or artificially shift the diffraction peak typically at the lowest or highest angles. This type of plot is generally linear at all angles except where the shielding occurs (Figure 6). This problem can be eliminated by paying close attention to proper sample positioning, checking the extreme ψ angles for shielding, and eliminating angles where shielding occurs.

USING POSITIVE AND NEGATIVE ψ ANGLES - There are essentially three types of nonlinear d-spacing versus $\sin^2\psi$ plots found when using both negative and positive ψ angles. The first type indicates the presence of shear stresses. In other words, a significant σ_{13} or σ_{23} component exists. In this instance the d-spacing versus $\sin^2\psi$ plot forms a "ψ split", i.e., the negative ψ angles form one slope while the positive ψ angles form a different slope.[1] To confirm the presence of shear stresses, one simply rotates the component 180° and repeats the measurement at the same location as the first measurement. If the negative and positive ψ angles switch position when the 180° rotation measurement is made, then shear stresses exist (Figure 7). Shear stresses are often seen in components that have been ground.

Two other types of ψ splits can be seen, neither of which are due to shear stresses. On samples with complex geometries, a ψ misset can produce a split. In this example the measurement location is not positioned at the center of rotation of the diffractometer or the sample is tilted such that a true $\psi = 0$ position is not achieved. Once again it is prudent to rotate the component 180° and repeat the measurement. If the sample was not aligned correctly, the negative and positive ψ angles do not switch position as they do when shear stresses are present (Figure 8). The remedy for this problem is to correctly align the sample and the diffractometer.

The last type of ψ split is due to focusing circle error and is related to the geometry of the diffractometer. The diffractometer used by these authors has a radius of about 200 mm. When stress measurements are made at angles less than 140° 2θ for this diffractometer radius, the negative ψ angles are not used because they produce an artificial shift in peak position due to the focusing circle geometry.[4] Once again this condition can be distinguished from the presence of shear stresses by rotating the component 180° and determining if the negative and positive ψ angles switch places.

<u>Line Broadening</u> - Diffraction peak width can also be measured to give additional insight into component condition. The percentage cold working in a component and the hardness in steels can be found by measuring the diffraction peak widths in calibration samples and comparing them to the peak widths in test samples.[5] As cold working or hardness increases, so does the width of the diffraction peak. Processes such as machining, grinding, and shot peening result in plastic deformation of a component's surface.

The effective depth of shot peening has been determined by not only monitoring the residual stress versus depth but by also plotting diffraction peak width versus depth (Figure 9).[6]

CONCLUSION

The x-ray diffraction technique is a powerful tool for measuring residual stress and additionally determining the metallurgical and processing condition of a component. Preferred orientation, large grain size, correct testing set up, and amount of cold working can be determined from the d-spacing versus $\sin^2\psi$ plots, integrated intensity versus $\sin^2\psi$ plots, and diffraction peak width.

REFERENCES

1. NOYAN, I.C. and COHEN, J.B., <u>Residual Stress Measurement by Diffraction and Interpretation</u>, Springer-Verlag New York Inc., 1987.
2. CULLITY, B.D., <u>Elements of X-Ray Diffraction</u>, Second Edition, Addison-Wesley Publishing Company, Inc., 1978.
3. JAMES, M.R., "The Use of Oscillation on PSD-Based Instruments for X-Ray Measurement of Residual Stress," Exp. Mech., 27, June 1987, No. 2, pp. 164-167.
4. PARDUE, E.B., JAMES, M.R., and HENDRICKS, R.W., <u>Advances in X-Ray Analysis</u>, 1987.
5. "Diffraction Notes," Vol. 1, No. 1 and No. 2, Lambda Research Inc., Spring 1987.
6. PARDUE, E.B. and LOWERY, L.A., "X-Ray Diffraction Stress Measurements on Various Shot-Peened Components," I.I.T.T. Conference, Nov. 6-7, 1990, Hollywood, CA (to be published).

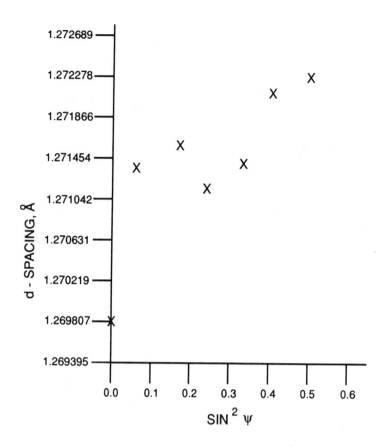

Fig. 1 - Undulating d-spacing versus $\sin^2 \psi$

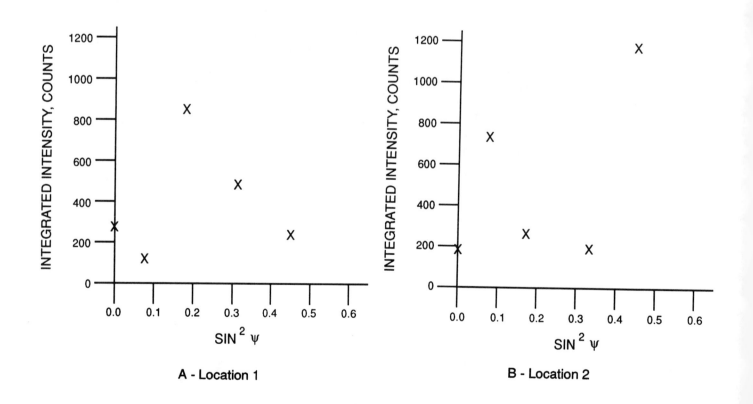

A - Location 1

B - Location 2

Fig. 2 - Large grain size indication

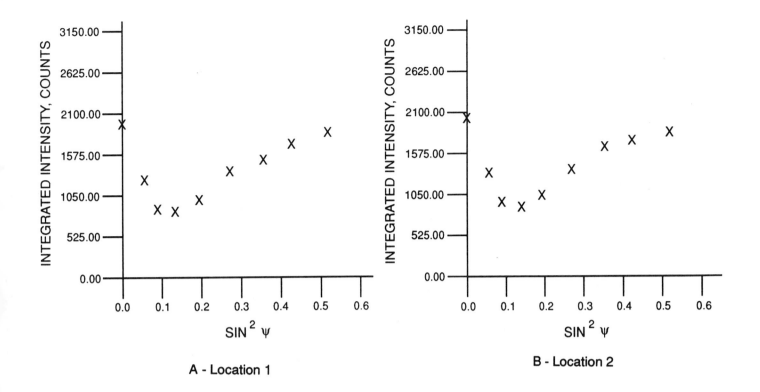

A - Location 1

B - Location 2

Fig. 3 - Preferred orientation indication

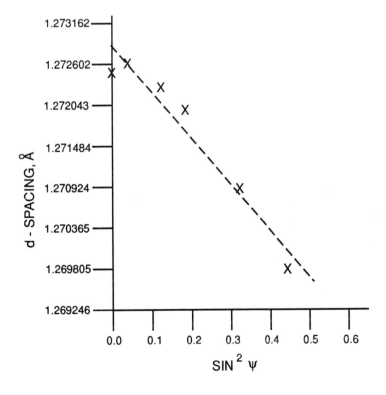

Fig. 4 - Stress gradient indication

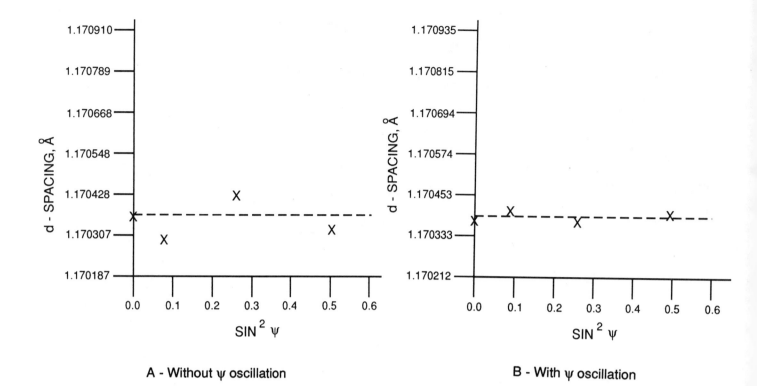

A - Without ψ oscillation

B - With ψ oscillation

Fig. 5 - Response due to $K_{\alpha 1}$ - $K_{\alpha 2}$ convolution and deconvolution

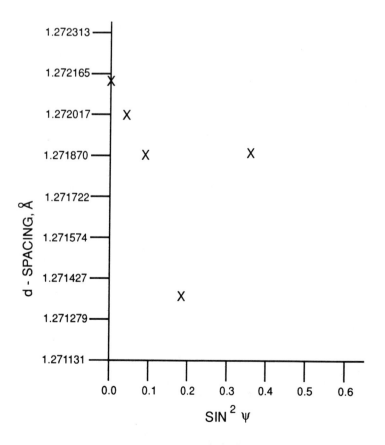

Fig. 6 - Shielding

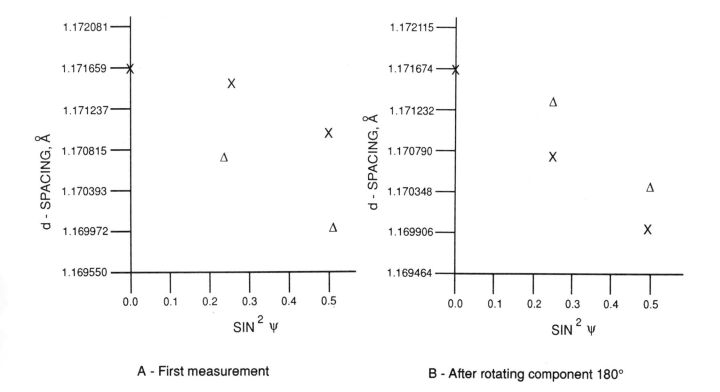

A - First measurement

B - After rotating component 180°

Fig. 7 - ψ split indicating shear stresses

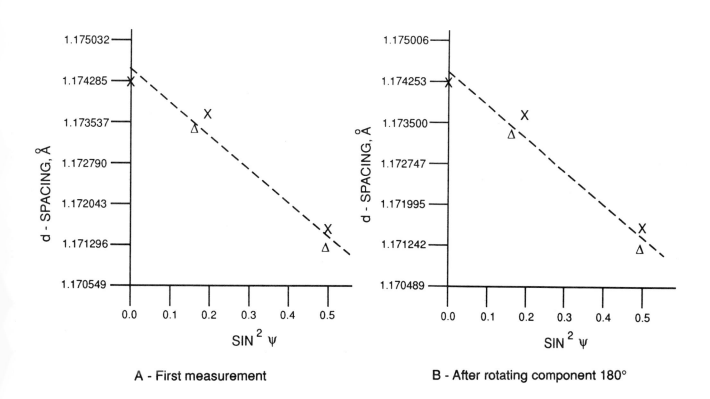

A - First measurement

B - After rotating component 180°

Fig. 8 - ψ misset due to improper alignment

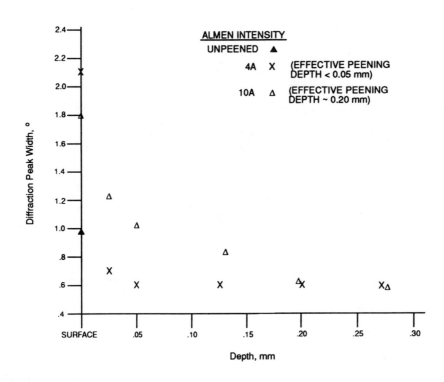

Fig. 9 - Effective shot peening depth determination from peak width data

Problems with Nondestructive Surface X-Ray Diffraction Residual Stress Measurement

P.S. Prevey
Lambda Research, Inc.
Cincinnati, Ohio

ABSTRACT

Because surface measurements are non-destructive, x-ray diffraction is often considered as a method of residual stress measurement for quality control testing. Unfortunately, errors caused by the presence of a subsurface stress gradient and difficulties in interpreting surface results often limit the usefulness of surface data. The magnitude of the potential errors, both in measurement and in interpretation, depends upon the nature of the subsurface residual stress distribution which can only be determined destructively. Although residual stress distributions subject to these problems are commonly encountered in practice, the question of the validity of non-destructive surface results is seldom adequately considered.

Examples are presented showing common residual stress distributions produced by grinding, nitriding and shot peening which are subject to errors in measurement and/or interpretation when measured only at the surface. The methods for determining the subsurface residual stress distributions and correction for penetration of the x-ray beam are discussed along with examples of their application. The need to determine the subsurface stress distribution in order to verify the accuracy of surface measurements is emphasized.

INTRODUCTION

X-ray diffraction (XRD) methods of residual stress measurement have been widely used for forty years, particularly in automotive and aerospace applications, and interest in the use of XRD stress measurement for quality control testing is increasing. Specifications now exist requiring that minimum levels of compression be achieved by shot peening, and limiting the tensile stresses allowed on EDM'd and ground surfaces. Commercial XRD residual stress measurement equipment, designed for both laboratory use and portable measurement in the field or shop environment, is readily available. However, there are problems with both measuring and interpreting XRD surface residual stress results which must be considered.

XRD provides an accurate and well established [1,2] method of determining the residual stress distributions produced by surface treatments such as machining, grinding and shot peening. XRD methods offer a number of advantages compared to the various mechanical, or the non-linear-elastic ultrasonic or magnetic methods currently available for the measurement of near-surface stresses. XRD methods are based upon linear elasticity, in which the residual stress in the material is calculated from the strain measured in the crystal lattice, and are not usually significantly affected by material properties such as hardness, degree of cold work or preferred orientation. XRD methods are capable of high spatial resolution, on the order of millimeters, and depth resolution, on the order of microns, and can be applied to a wide variety of sample geometries. The macroscopic residual stress and information related to the degree of cold working can be obtained simultaneously by XRD methods. XRD methods are applicable to most polycrystalline materials, metallic or ceramic, and are non-destructive at the sample surface.

The most common problems encountered in using XRD methods of residual stress measurement are related to the high precision required for measurement of the diffraction angles, which in turn require accurate sample/instrument alignment and precise methods of diffraction peak location [3]. XRD methods are applicable only to relatively fine-grained materials, and often cannot be applied to coarse-grained castings. The shallow depth of penetration of the x-ray beam can be a disadvantage when trying to characterize a subsurface stress distribution with only surface measurements. Rarely, extreme preferred orientation and sheer stresses at the sample surface cause errors.

This paper briefly describes the assumptions, theory and limitations of XRD residual stress measurement as applied to the study of residual stress distributions produced by such processes as machining, grinding and shot peening. Special mention is made of problems commonly encountered in both obtaining and interpreting surface data from such samples.

THEORY

Macroscopic Residual Stress Measurement

Because the depth of penetration of the x-ray beam is extremely shallow, the diffracting volume can be considered to represent a free surface under plane stress. As shown in Figure 1, the biaxial surface stress field is defined by the principal (residual and/or applied) stresses, σ_1 and σ_2, with no stress normal to the surface. The stress to be determined is the stress, σ_ϕ, lying in the plane of the surface at an angle, ϕ, to the maximum principal stress, σ_1. The direction of measurement is determined by the plane of diffraction. The stress in any direction (for any angle, ϕ) can be determined by rotating the specimen in the x-ray beam. If the stress is measured in at least three different directions, the principal stresses and their orientation can be calculated.

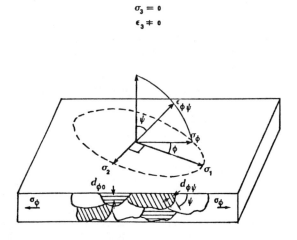

Fig. 1 Plane Stress at a Free Surface

Consider the strain vector, $\epsilon_{\phi\psi}$, lying in the plane defined by the surface normal and the stress, σ_ϕ, to be determined. $\epsilon_{\phi\psi}$ is at an angle ψ, to the surface normal, and can be expressed in terms of the stress of interest and the sum of the principal stresses as,

$$(1) \quad \epsilon_{\phi\psi} = \frac{1+\nu}{E} \sigma_\phi \sin^2\psi - \left(\frac{\nu}{E}\right)\left(\sigma_1 + \sigma_2\right)$$

The sample is assumed to consist of a large number of small grains or crystals, nominally randomly oriented, as shown schematically in Figure 1. The crystal lattice consists of planes of atoms identified by their Miller indices, (hkl). The spacing between a specific set of lattice planes, for example, the (211) planes in a steel, will be equal regardless of the orientation of the lattice planes relative to the sample surface in a stress-free specimen. The lattice spacing will be expanded or compressed elastically (by an amount dependent upon the orientation of the lattice planes) by any stress present in the specimen. The state of stress within the depth of penetration of the x-ray beam can be determined by measuring the lattice spacing at different orientations to the sample surface.

The only crystals which diffract x-rays are those which are properly oriented relative to the x-ray beam to satisfy Bragg's Law,

$$(2) \quad n\lambda = 2d\sin\theta$$

where λ is the known x-ray wavelength, n is an integer (typically 1), θ is the diffraction angle, and d is the lattice spacing. XRD can be used to selectively measure the lattice spacing of only those crystals of a selected phase which have a specific orientation relative to the sample surface by measuring θ and calculating d from Equation 2.

The lattice spacing can be determined for any orientation, ψ, relative to the sample surface by merely rotating the specimen. If σ_ϕ is a tensile stress, the spacing between lattice planes parallel to the surface will be reduced by a Poisson's ratio contraction, while the spacing of planes tilted into the direction of the tensile stress will be expanded. If we express the strain in terms of the crystal lattice spacing,

$$(3) \quad \epsilon_{\phi\psi} = \frac{d_{\phi\psi} - d_o}{d_o}$$

where d_o is the stress-free lattice spacing and $d(\phi,\psi)$ is the lattice spacing measured in the direction defined by ϕ and ψ. By substituting Equation 3 into Equation 1, the lattice spacing measured in any orientation can be expressed as a function of the stresses present in the sample and the elastic constants in the (hkl) crystallographic direction used for stress measurement,

$$d(\phi,\psi) = \left(\frac{1+\nu}{E}\right)_{hkl} \sigma_\phi d_o \sin^2\psi - \left(\frac{\nu}{E}\right) d_o (\sigma_1 + \sigma_2) + d_o \quad (4)$$

It should be noted that the elastic constants in the (hkl) direction may differ significantly from the values obtained by mechanical testing because of elastic anisotropy, and should be determined empirically. [4]

Examination of Equation 4 shows that for the plane-stress model assumed, the lattice spacing measured at any angle, ψ, to the surface normal will vary linearly as a function of $\sin^2\psi$. The actual lattice spacing of the (311) planes plotted as a function of $\sin^2\psi$ for shot peened 5056 aluminum is shown in Figure 2. The intercept of the plot is equal to the unstressed lattice spacing, d_o, minus the Poisson's ratio contraction caused by the sum of the principal stresses. Because the value of the lattice spacing measured at $\psi = 0$ differs by not more than 0.1 percent from the stress-free lattice spacing, the intercept can be substituted for d_o. The stress is determined from the slope, the elastic constants and the value of d measured at $\psi = 0$. The residual stress can then be calculated without reference to a stress-free standard.

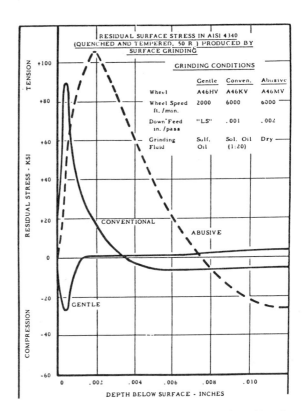

Fig. 3 Subsurface Stress Distributions Produced by Diverse Grinding Conditions in 4340 Steel [6]

XRD macroscopic residual stress measurement yields the arithmetic average stress in a diffracting volume defined by the dimensions of the irradiated area and the depth of penetration of the x-ray beam. The residual stress in that volume is assumed to be constant both in the plane parallel to the surface and as a function of depth. Unfortunately, the stress distributions encountered in many samples of practical interest violate these assumptions, especially at the surface where measurements may be performed non-destructively.

PROBLEMS WITH SURFACE MEASUREMENT

There are three primary difficulties associated with both obtaining and interpreting surface x-ray diffraction residual stress results. First, the surface residual stresses present on many samples of practical interest simply are not representative of the processes which produced them. Second, many machining and grinding practices produce variations in the surface residual stresses which are so large that surface results are of little value. Third, many material removal and surface treatment processes produce subsurface stress distributions which vary significantly within the depth of penetration of the x-ray beam, and can cause significant experimental error in the measurement of the surface stress.

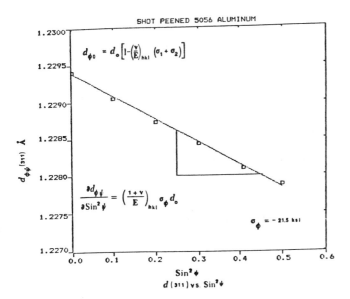

Fig. 2 Linear Dependence of Lattice Spacing With $\sin^2\psi$ in Shot Peened Aluminum

Surface Stresses May Not be Representative

Many of the processes of common interest, such as grinding, shot peening, nitriding, etc., can produce nearly identical surface residual stresses for a wide range of processing variables. This feature of the stress distribution may prohibit the use of non-destructive surface residual stress measurements, regardless of measurement accuracy, from being useful for quality control testing.

In the case of grinding, where x-ray diffraction is frequently considered as a means of detecting tensile stresses, the surface stress may be nearly independent of the grinding parameters. Figure 3 shows three classic representations of gentle, conventional and abusive grinding of 4340 steel measured by a mechanical technique of layer removal and stress relaxation. The near-surface residual stresses range from only 0 to 140 MPa for an extreme range of grinding conditions. Similar surface stresses produced by completely different surface treatments are commonly revealed by x-ray diffraction, as in Figure 6.

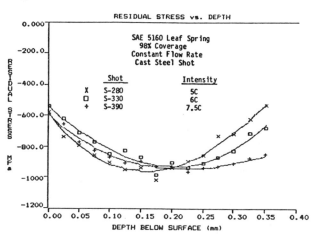

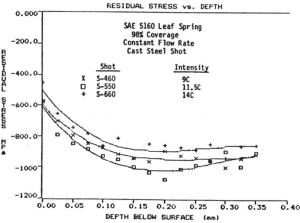

Fig. 4 Subsurface Stress Distributions Produced by Shot Peening SAE 5160 Steel, Showing Similar Surface Values [7]

Shot peening also frequently produces nearly identical surface residual stresses for a wide variation in peening parameters, including shot size and Almen intensity. Figure 4 shows results for 5160 steel leaf springs shot peened from a 5C to 14C intensity with shot sizes ranging from S-280 to S-660. The surface residual stresses are virtually identical for all six peening methods, although significant differences are observed in the depth of the peened layer. Figure 5 compares the stress distributions produced by shot peening Inconel 718 to 6-8A and 5-7C intensities. The results near the surface are, again, virtually identical, but there is a pronounced variation in the depth of the compressive layers. Similar surface results are observed on shot peened 8620 steel gears as well, even though the fatigue life is well correlated to the depth of the peened layer. Figure 6 shows comparable surface residual stresses in carburized 8620 steel produced by grinding and shot peening to an 18A intensity. Non-destructive surface residual stress measurement could not be used to distinguish whether the part was in the ground or shot peened condition. A variety of other cold abrasive processes such as sand or grit blasting, wire brushing and even polishing with abrasive paper will produce surface residual stresses indistinguishable to those achieved by shot peening.

A given level of surface residual stress is a necessary but not a sufficient condition to indicate that a critical component may have been correctly processed. The surface residual stress measured non-destructively by x-ray diffraction, or any other means, is frequently inadequate for process control testing.

Surface Stress Variation

Many metal removal processes, particularly those involving chip formation such as machining and grinding, can generate pronounced local fluctuations in the surface residual stress. Variation in the depth and magnitude of the deformed layer and the heat input near the surface during chip formation can result in large differences in the resulting surface residual stresses over distances on the order of millimeters.

The apparent surface residual stress measured by x-ray diffraction will then be dependent upon both the size and the positioning of the irradiated area used for measurement. If a small irradiated area is used, the assumption of uniform stress within the irradiated area may be satisfied, and the stress variation at the sample surface will be revealed. The surface stress variation can be so pronounced as to render non-destructive measurement useless for process control.

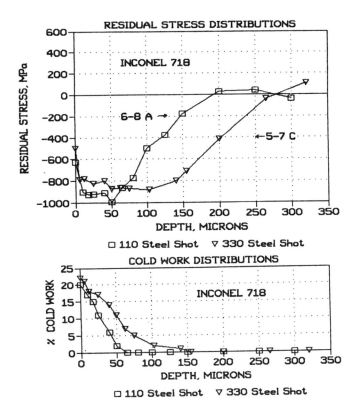

Fig. 5 Variation in Depth of the Stress Distributions Produced in Shot Peened Inconel 718, Showing Similar Surface Results [8]

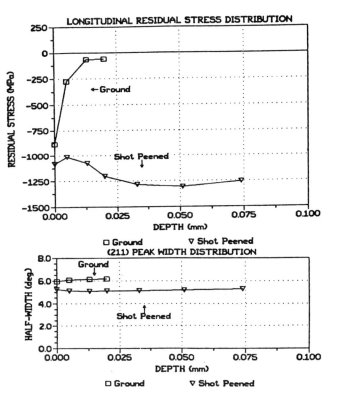

Fig. 6 Residual Stress and Peak Width Distributions Produced by Shot Peening (18A) and Grinding of Carburized 8620 Steel

Alternately, the irradiated area may be made large enough to provide a useful average surface stress, but then the assumption of uniform stress in the irradiated area may be violated. The surface stress measured will be the arithmetic average within the irradiated area, and will be dependent upon the details of technique such as the ψ angles used during measurement.

Figure 7 shows the surface residual stress measured using an irradiated area of 12mm x 0.5mm across a 19mm wide surface of a ground 4340 steel sample. The surface stresses vary by nearly 600 MPa from a region of maximum compression near one edge of the sample to maximum tension in a burned area. The use of a larger irradiated area, plotted as a line through the individual data points, yields the arithmetic mean.

Subsurface measurements at the points of minimum and maximum surface stress shown in Figure 8 reveal subsurface tension at both locations. Comparable variations in the surface residual stress are seen in Figure 9 for milled Inconel 718. The stress variation is greatest at the sample surface. Extreme local variation of the surface stress often encountered on the machined and ground

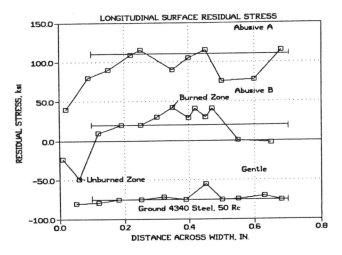

Fig. 7 Variation in Surface Residual Stress Across the Surface of Ground 4340 Steel [9]

samples, may prohibit the use of x-ray diffraction residual stress measurement for quality control testing.

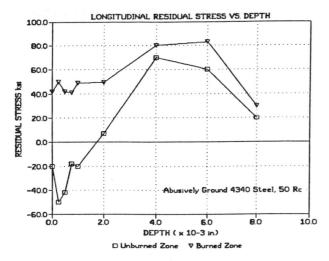

Fig. 8 Subsurface Stress Distributions in Adjacent Areas of Ground 4340 Steel Showing Surface Stress Variation and Similar Subsurface Peak Stress [9]

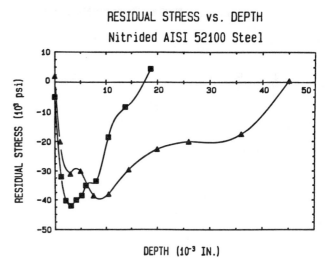

Fig. 10 Subsurface Stress Distributions Produced by Nitriding AISI 52100 Steel, Showing Pronounced Near-Surface Stress Gradient [10]

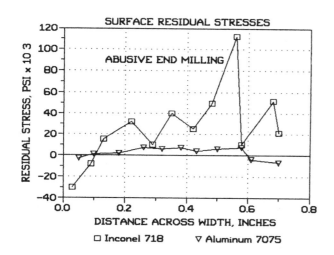

Fig. 9 Surface Stress Variation in End Milling of Inconel 718 and Aluminum [9]

Errors Due to Subsurface Residual Stress Gradients

For most materials of practical interest and the radiations used for residual stress measurement, the effective depth of penetration of the x-ray beam is quite shallow. Nominally 50% of the diffracted radiation originates from a depth of less than 10 μm. However, the x-ray beam is attenuated exponentially as a function of depth. The rate of attenuation is governed by the linear absorption coefficient, which depends upon the composition and density of the specimen and the radiation used.

Any "surface" measurement is, therefore, actually an exponentially weighted average of the stress at the surface and in the layers immediately beneath it. As noted in the theory section, the assumption was made that the residual stress is constant throughout the depth of penetration of the x-ray beam. Unfortunately, for many samples of practical interest, the stress varies rapidly with depth beneath the surface, and the assumption of constant stress is violated. The result can be errors as large as 600 MPa.

The sign and magnitude of the potential error is dependent upon the subsurface stress gradient; i.e., the direction and rate of change of stress with depth into the sample surface. Because the depth of penetration of the x-ray beam also varies with the angles ψ, and 2θ, the apparent surface residual stress will depend upon the details of the technique chosen, specifically the radiation and ψ angles selected, if a significant subsurface stress gradient exists.

Figure 10 shows examples of large subsurface stress gradients produced by two different methods of nitriding 52100 steel. The grinding stress distributions shown in Figure 3 show large stress gradients at the surface, both positive and negative. Figure 5 shows a pronounced gradient in the "hook" commonly seen at the surface of shot peening stress distributions. Figure 11 depicts a complete reversal of the stress distribution within 50 microns of the surface observed on abrasively cut Inconel 718.

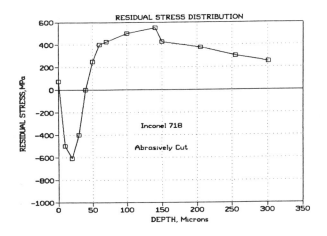

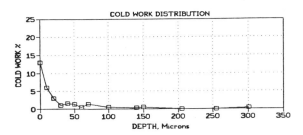

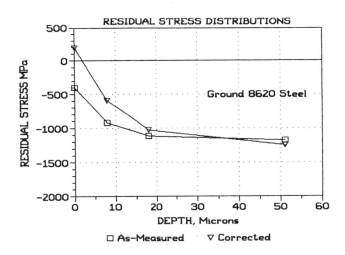

Fig. 13 Effect of Correction for Penetration of the X-ray Radiation into a Subsurface Stress Gradient, Showing a Change of Sign at the Surface

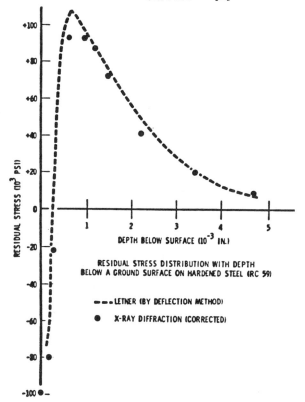

Fig. 11 Subsurface Stress Distributions in Abrasively Cut Inconel 718, Showing Complete Stress Reversal Near the Surface [8]

Fig. 12 Subsurface Stress Distribution in Ground Steel Measured by Mechanical and X-Ray Diffraction Methods with Correction for the Near-Surface Stress Gradient [11]

It is possible to correct for the errors caused by the penetration of the x-ray beam into the stress gradient, provided subsurface measurements are made by electropolishing to remove layers with sufficient depth resolution to accurately establish the stress gradient. Koistinen and Marburger [5] developed a method of calculating the true residual stress by unfolding the exponential weighting caused by penetration of the x-ray beam. Their often cited example of agreement between x-ray diffraction and mechanical methods of residual stress measurement in ground steel, reproduced in Figure 12, shows agreement only because the correction was applied. The figure is reproduced exactly as it appears in their original publication.

Figures 13 and 14 show positive and negative corrections, respectively. As seen in Figure 13, the uncorrected surface stress may even be of the wrong sign.

Non-destructive surface residual stress measurements cannot be corrected for errors caused by penetration of the x-ray beam into a varying stress field. Therefore, surface results must be interpreted with caution. The true surface stress frequently cannot be accurately determined by surface measurement alone.

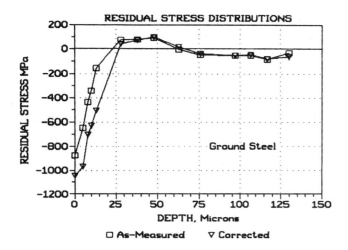

Fig. 14 Effect of Correction for Penetration of the X-ray Radiation into a Stress Gradient in Ground Steel

CONCLUSIONS

The limitations inherent in the use of surface x-ray diffraction residual stress measurements have been shown to result in three areas of concern, which must be considered before non-destructive surface results may be used reliably.

First, there frequently is no correlation between the surface residual stress and the method of processing which produced the stress distribution. Subsurface stresses often differ significantly from the surface value.

Second, the surface stresses produced by many material removal processes, particularly machining and grinding, will often vary significantly over short distances. The surface stress measured is then dependent upon the details of the measurement technique, such as the irradiated area size and positioning.

Third, many processes of practical interest result in a rapid change in the residual stress immediately beneath the surface, within the depth of penetration of the x-ray beam. This results in errors which can approach 600 MPa and even alter the sign of the apparent results. The effects of penetration of the x-ray beam can only be corrected if subsurface results are obtained.

REFERENCES:

1. HILLEY, M.E., ed., "Residual Stress Measurement by X-ray Diffraction," SAE J784a, Society of Automotive Engineers, Warrendale, PA (1971).

2. ASTM, "Standard Method for Verifying the Alignment of X-ray Diffraction Instrumentation for Residual Stress Measurement," E915, Vol. 3.01, Philadelphia, PA, 809-812 (1984).

3. PREVEY, P.S., "The Use of Pearson VII Distribution Functions in X-ray Diffraction Residual Stress Measurement," ADV. IN X-RAY ANALYSIS, Vol. 29, Plenum Press, NY, 103-112 (1986).

4. PREVEY, P.S., "A Method of Determining the Elastic Properties of Alloys in Selected Crystallograhic Directions for X-ray Diffraction Residual Stress Measurement," ADV. IN X-RAY ANALYSIS, Vol. 20, Plenum Press, NY, 345-354 (1977).

5. KOISTENEN, D.P., and MARBURGER, R.E., Trans. ASM, Vol. 51, 537 (1959).

6. KOSTER, W.P. et al, "Surface Integrity of Machined Structural Components," US Air Force Materials Laboratory Technical Report No. 70-11 p. 112, 1970.

7. DeLITIZIA, A.T., "Influence of Shot Peening on the Residual Stresses in Spring Steel Plate," p. 237-240, Second International Conference on Shot Peening, 1984.

8. PREVEY, P.S., "The Measurement of Subsurface Residual Stress and Cold Work Distributions in Nickel Base Alloys," Residual Stress in Design, Process and Materials Selection, ASM International, pp. 11-19, 1987.

9. PREVEY, P.S. and Field, M., Annals of the CIRP, Vol.24, 1, 1975, p.498-500.

10. KOISTINEN & MARBURGER, Trans. ASM, Vol. 67, 1964.

11. KOISTINEN & MARBURGER, Trans. ASM, Vol. 51, p. 537, 1959.

Evaluation of the Stress Distribution in Welded Steel by Measurement of the Barkhausen Noise Level

K. Tiitto
American Stress Technologies, Inc.
Pittsburgh, Pennsylvania

A.S. Wojtas, W.J.P. Vink, and G. denOuden
Delft University of Technology
Delft, The Netherlands

ABSTRACT

This paper presents the results of residual stress measurements on welded 6 mm thick low carbon steel plate by using the Barkhausen noise analysis.

To convert the Barkhausen noise signal into stress units, both uniaxial and biaxial calibration was carried out on test samples cut from the same steel plate as used for the welded samples. The calibration samples were bent to generate different combinations of stresses/strains in both longitudinal and transverse directions of the sample, and these strains were measured with strain gages. Uniaxial calibration curves and biaxial calibration surfaces, depicting Barkhausen noise as a function of strain, were plotted and subsequently used to generate the stress data for the welded samples.

A comparison was made between the stresses obtained with the Barkhausen noise method and X-ray diffraction. A good correlation was found, provided that proper care was taken when preparing the samples.

Additional tests were made to determine the accuracy of the method from the point of view of the instrumentation.

On the basis of the overall results obtained, it appears that the Barkhausen noise analysis offers a reliable method to evaluate stresses in the vicinity of welds.

INTRODUCTION

Phase transformations and thermal contraction during and directly after arc welding can create significant residual stresses in and around a weld. These stresses may impair the mechanical properties and for this reason knowledge of these stresses is important. Techniques applied to measure residual stress include x-ray diffraction and blind hole drilling. Application of either of these methods, however, is in general time consuming and often difficult under practical conditions.

In the case of ferromagnetic materials, residual stresses can also be measured by analysis of the Barkhausen noise which is generated by external magnetization /1, 2, 3/. In magnetostrictive materials such as ferritic steels, the Barkhausen noise increases or decreases under the influence of a tensile or compressive stress, respectively, applied in a direction parallel to that of the magnetic field. Technological developments in recent years have made it possible to construct equipment capable of measuring the changes in the Barkhausen noise with high accuracy.

Aim of the study presented in this paper was to examine the feasibility of the Barkhausen noise analysis method for the evaluation of residual stresses in welded steel. The results were compared with those obtained by x-ray diffraction, which is a well-established method to determine residual stress.

PRINCIPLE OF MEASUREMENT

The magnetoelastic Barkhausen noise method is based on the concept of ferromagnetic domains discovered in 1919 by Professor Barkhausen at the University of Dresden, Germany.

Ferromagnetic materials consist of microscopic, magnetically ordered regions called domains. Each domain, resembling a bar magnet, is magnetized along a certain crystallographic "easy direction" of magnetization. Domains are separated from one another by walls within which the direction of magnetization usually turns 180° or 90°. The net magnetization of a material is the average of the magnetizations within all domains. When a magnetic field or mechanical stress is applied to a ferromagnetic material, changes take place in its domain structure by abrupt movement of domain walls or rotation of domain magnetization vectors. These changes, in turn, produce changes in the overall sample magnetization, as well as in sample dimensions.

If a coil of conducting wire is placed near the sample while a domain wall moves, the resulting change in magnetization will induce an electrical pulse in the coil. The first observations of domain wall motion based on electromagnetic induction were made by Barkhausen /1/. He discovered the magnetization process, i.e., the hysteresis curve is not continuous but consists of small steps generated when domains abruptly jump from one position to another under an applied magnetic field. When the electrical pulses produced by all domain movements under the coil area are added together, a noise-like signal called Barkhausen noise is generated.

Barkhausen noise has a power spectrum starting from the magnetizing frequency and extending up to 250-500 kHz /4/. The noise is exponentially damped as a function of distance inside the material. This is primarily due to the eddy current damping of the electromagnetic fields generated by moving domain walls /5, 6/. As these fields can be measured on the surface only, the extent of damping determines the depth from which information can be obtained. The main factors affecting this depth are the frequency range selected for signal analysis and the conductivity and permeability of the test material. Measurement depths for practical applications in steels vary between 0.01 and 3.0 mm.

Intensity of Barkhausen noise depends on stresses and microstructure of the material. Properly calibrated, Barkhausen noise can be used for rapid nondestructive testing of uniaxial and biaxial surface stresses. The ultimate accuracy of the method depends on how closely the microstructural parameters of the calibration sample and actual test piece agree.

EXPERIMENTAL

MATERIALS - Low carbon steel (Fe 510 according to ISO specification) plates with thickness of 6 mm were used in the as-received condition. Samples with dimensions of 200 mm x 300 mm were cut out of these plates. The surface oxide layer was removed using flat bench grinder. in order to minimize the induced grinding stresses, grit 240, 500, 800 and 1200 sandpaper was used to grind the surface. In all cases the roughness did not exceed 0.2 mm.

WELDING PROCEDURE - Full penetration welds were produced in the samples by means of autogeneous GTA welding. In all cases a 4.8 mm thoria doped tungsten electrode was used. During welding the arc length was maintained at 4 mm. Shielding gas used was 99.5% argon. Welding was carried out at a travel speed of 3 mm/s, a welding current of 218 A and an arc voltage of 14 V.

BARKHAUSEN NOISE MEASUREMENTS - The samples welded were measured with ROLLSCAN and STRESSCAN central units and a narrow miniature general purpose sensor. A grid was drawn on the surface of each sample so that the grid lines were at a distance of 2.5 mm from each other up to 20 mm away from the weld, and beyond this distance the grid lines were separated by 5 mm from each other. Both longitudinal and transverse (to the weld) Barkhausen noise measurements were made at the grid nodes.

The nominal depth of measurement was 0.02 mm, which was calculated from the damping of the Barkhausen noise for the frequency range of 70-200 kHz used in this work.

CALIBRATION PROCEDURE - One-dimensional (uniaxial) calibration has been traditionally used for magneto-elastic Barkhausen noise (BN) measurements. Recently, a more accurate two-dimensional (biaxial) procedure has been developed to cover the testing of a more complex biaxial stress condition /7/. The biaxial procedure was used in this work. However, both uniaxial and biaxial procedures are described below.

BN level is expressed here in terms of MP, magnetoelastic parameter (measured in millivolts), proportional to the RMS level of Barkhausen noise after amplification and filtering.

<u>Uniaxial Calibration</u> - Uniaxial calibration is performed by measuring the BN level as a function of compression and tension in one (axial) direction of the test piece only. This procedure is useful whenever the transverse biaxial component of stress is less than 25% of the elasticity limit of the material. Stresses can be varied by a bending test with strain gages, cantilever beam test with known loads or tensile test. The first method, being considered the most reliable one, is described here in more detail.

A typical calibration sample is a flat bar, approximately 6 mm thick with a strain gage mounted next to the location where MP is to be measured. Care should be taken not to bend the sample beyond its elasticity limits as the resulting plastic deformation will change the residual stresses in the bar and complicate the interpretation of test results.

The uniaxial calibration procedure consists of the following main steps:

(i) Determine separately the elasticity limits in tension (ε_t) and compression (ε_c) by increasing the maximum bending load in steps and releasing it again until the bending cycle becomes irreversible, i.e., MP and strain readings do not return to their initial "no load" values. Record MP and strain (ε) at each step.

(ii) Plot the curve MP vs. ε within the elasticity limits from the above data. Set $\varepsilon = 0$ for no load strain to represent the apparent zero strain point on the curve.

(iii) Determine the true zero strain point $\varepsilon = \varepsilon_o$. Assuming the material yields symmetrically in tension and compression, this is the middle point between the elasticity limits, or

$$\varepsilon_o = 1/2\,(\varepsilon_t - \varepsilon_c)$$

The MP reading corresponding to the true zero

strain, MP_o, is the projection of ε_o on the MP axis. The difference between apparent and true zero strains, $\varepsilon(R)$, represents the one-dimensional residual strain in the calibration bar.

Accordingly, the complete calibration curve, elasticity limits, zero stress/strain point and residual stress/strain present in the sample are determined by uniaxial calibration (strains can be converted into stresses by Young's modulus and Poisson's ratio).

<u>Biaxial Calibration</u> - It has been shown that the transverse stress will have an effect on uniaxial calibration curves /8, 9, 10/, and a significant error in stress evaluation may be involved if uniaxial calibration is applied to test a strongly biaxial stress condition. This error can be greatly reduced with the application of biaxial calibration.

Basically, biaxial calibration consists of determining a set of uniaxial calibration curves by varying the transverse stress so as to cover all combinations of longitudinal and transverse stress (principal stress) components. At each combination, MP values parallel to both stress components are measured, i.e., the MP combination corresponding to each principal stress combination is determined. This expands the one-dimensional calibration curve into a fairly complex dual calibration surface. Biaxial calibration is a significant improvement in the state-of-the-art, as it allows for separating anisotropy from stress and building the anisotropy correction into the calibration data.

Flat cross shaped test pieces are used for biaxial calibration instead of a flat bar. With a motorized mechanical fixture, all combinations of a biaxial strain field are produced by independently bending both legs of the cross in tension and compression as if they were separate slabs. MP is measured parallel to each leg at the center of the cross. One leg is designated L (longitudinal; rolling direction); the other T (transverse). Longitudinal and transverse strains are measured by miniature strain gages mounted on each bending axis next to the MP test location.

To evaluate the calibration data, steps (i) to (iii) as per uniaxial calibration are first performed with respect to both L and T legs and a biaxial zero strain condition of the sample determined. Next, the sample is bent in the fixture into biaxial zero strain, whereafter strains L and T are systematically varied relative to each other and corresponding BN readings MP(L) and MP(T) measured. The complete calibration data obtained in this way can be arranged into a table or can be presented as a three dimensional graphic illustration of the calibration surface.

Error may be involved when testing biaxial stresses in strongly anisotropic materials if (i) such directions of anisotropy are not known in actual test components, or (ii) principal stress directions in actual test components differ significantly from L and T directions.

Extensive calibration tests have shown that the BN response extends over the full elastic range in most steels, with some leveling off characteristic of the S shape at the tails of the calibration curve. Hard martensitic steels seem to be an exception, with early saturation being noticed in the compressive end of this curve.

RESULTS AND DISCUSSION

WELDING STRESS - The MP values measured on the grid nodes of the welded Fe 510 plates are presented in Figure 1a for the longitudinal direction and in Figure 1b for the transverse direction. These figures were obtained by using a computer graphics software.

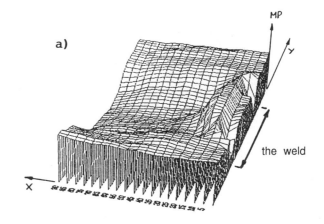

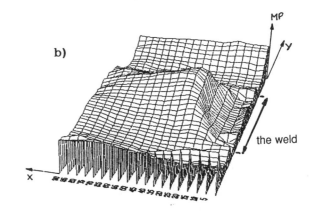

Fig. 1 - Three-dimensional presentation of measured MP values with magnetization direction (a) parallel and (b) perpendicular to the weld seam.

In order to convert MP values into stress, calibration surface was generated as described above, see Figure 2. With the aid of this calibration surface, the measured MP values were converted into stress as shown in Figure 3.

It can be seen in Figures 1 and 3 that the longitudinal stress σ_L is tensile within a short range from the center of the weld and becomes compressive at an increasing distance. In the weld and in the adjacent zone, the longitudinal stress is close to zero or has a low positive value. The transverse stress σ_T is compressive within the closest proximity of the weld but rises sharply to reach a tensile peak and decreases slowly, remaining tensile. The exact locations of stress peaks and the transitions between the compressive and the tensile stresses depend on the welding parameters and the properties of the steel. The stress distributions obtained from both samples discussed here exhibit a good agreement with results given in literature /11-14/.

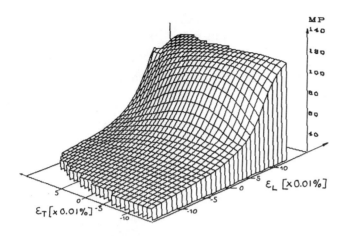

Fig. 2 - A three-dimensional presentation of the biaxial calibration surface.

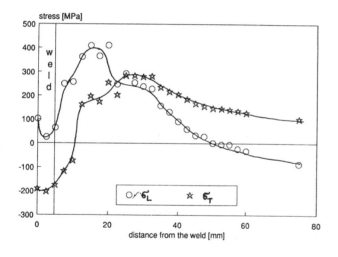

Fig. 3 - Stress profiles as measured in the directions parallel (σ_L) and perpendicular (σ_T) to the weld.

An extensive investigation of the stresses in and around the weld was carried out by Brand et al /15/ at Delft University of Technology. A set of samples welded with different heat input levels were studied by x-ray diffraction using the $\sin^2\psi$ method. Longitudinal and transverse stress profiles as a function of distance from the weld were generated. These results are compiled in Figure 4 together with Barkhausen noise results obtained in this work for the same sample. A comparison of the results obtained in this research with those acquired by Brand yields interesting observations. Qualitatively, there is a good correlation between the two sets of data. High stresses are found close to the seam especially in the longitudinal direction and these

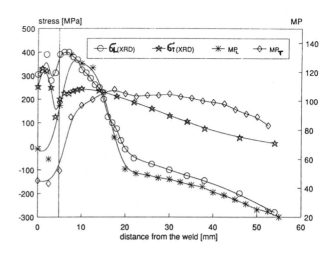

Fig. 4 - Comparison between MP values obtained in this work and x-ray residual stress measured by Brand /15/.

stresses are changed to compressive stresses further away from the weld seam. The correlation between transverse measurements seems reasonable in the base metal. However, there is a substantial difference between Barkhausen noise values and stress measured by x-ray diffraction in the weld. The reason for this difference is presently under study.

ACCURACY - The accuracy of the Barkhausen noise method can be characterized as depending on three following aspects:

1. The slope of the calibration curve.

    A unit change in the measured MP level will yield a change in strain inversely proportional to the tangent of the slope of the curve in a given point. The slope of the calibration curve obtained in this research varies between 2 and 8 MP per 100$\mu\epsilon$ at the elasticity limits and in the vicinity of the zero strain point respectively.

2. The fluctuation in time of the measured MP values.

    Due to the statistical nature of the Barkhausen noise, the measured MP value will oscillate slightly in time. The fluctuations observed here were proportional to the magnitude of the measured MP. The possible error introduced by such variations has been estimated to be less than 0.5%.

3. The repeatability of the measurement.

    Measurements of a stress profile repeated after a month were found to be within 2% of the initial values.

The additive character of the above sources of inaccuracy leads to the conclusion that the overall accuracy of the method for most practical cases will remain within 4-5%, not including the error due to possible variation in material parameters.

CONCLUSIONS

In this report residual stresses in a welded Fe 510 steel plate were measured by Barkhausen noise analysis. The following conclusions can be drawn:

1. An increase in imposed stress in the direction parallel to that of the magnetic field lines from compression through tension, causes an increase in measured Barkhausen noise intensity (MP). It has been also observed that changes in stress in the direction perpendicular to that of sample magnetization alter measured MP values. In order to evaluate residual stresses in steel from measurement of the Barkhausen noise, calibration is necessary using a sample made of the same material.

2. Comparative measurements of residual stress using Barkhausen noise analysis and x-ray diffraction have shown a good correlation between these two methods. It is essential for both techniques that sample surfaces are properly treated prior to the measurements. Comparison between results of measurements obtained using uniaxial and biaxial calibrations shows that a substantial error can be made using uniaxial calibration data to convert complex profiles of longitudinal and transverse stresses.

3. High speed of measurement and uncomplicated handling of the equipment allow extensive evaluation of residual stress distributions and detection of principal stress directions on large surfaces.

4. This technique applied in nondestructive testing is capable of dynamic sensing and is suitable for automated as well as for manual inspection and process control.

5. As this test has shown, Barkhausen noise analysis offers a fast and accurate method for stress measurement. It is recommended that further experiments be set up to examine the influence of changing microstructure as it occurs in the weld and to establish the accuracy in such circumstances.

REFERENCES

1. Barkhausen, H., *Phys. Zeitschrif,* **20**, 401-403 (1919)

2. Gerlach, H. and P. Lertes, *Phys. Zeitschrift,* **22**, 568 (1921)

3. Zschiesche, K., *Phys. Zeitschrift,* **23**, 201 (1922)

4. Tiitto, S., Saynajakangas, S. and R. Rulka, IEEE Trans. Magn., **12**, 406 (1976)

5. Tiitto, S. and S. Saynajakangas, *Spectral Damping in Barkhausen Noise*, IEEE Trans. Magn., **11**, 1666 (1975)

6. Namkung, M., Utrata, D., Heyman, J. S. and S. G. Allison, "Solid Mechanics Research for Quantitative Nondestructive Evaluation", pp. 301-318, Martinus Nijhoff Publishers, Dordrecht, The Netherlands, (1987)

7. EX: Tiitto, S., U.S. Patent Application No. 332,478

8. Karjalainen, P. and R. Rautioaho, *Detection of Fabrication Stresses by the Barkhausen Noise Method,* Conference on Effects of Fabrication Related Stresses on Product Manufacture and Performance, The Welding Institute, Abington Hall, Cambridge, 13-1 (1985)

9. Furuya, T., Shimada, H. and Y. Ito, Journal of JSNDI, 36(8), 530 (1987) (in Japanese)

10. Loomis, K., "Barkhausen Biaxial Stress/Strain Measurement System", Proceedings of the 13th International Nondestructive Testing Conference, San Antonio, Texas (1981)

11. Tietz, H. D., *Grundlagen der Eigenspannungen,* D.V.G., Leipzig (1982)

12. Wohlfahrt, H., "Schweisseneigenspannungen", DHTM **31**, 56-71 (1976)

13. Wohlfahrt, H., "Die Bedeutung der Austenietumwandlung fur die Eigenspannungsentstehung beim Schweissen", HTM **41**, 248-257 (1986)

14. Willemse, P. F. and Kolster, B. H., Lastechniek, **52**, 4-7 (1986)

15. Brand, P. C., *Restspanningen bij het lassen van staal,* Delft University of Technology, (1987)

Practical Applications of Residual Stress Technology, Conference Proceedings, Indianapolis, Indiana, USA, 15-17 May 1991

Ultrasonic Measurements of Residual Stress in Railroad Wheels

R.E. Schramm, A.V. Clark, D.V. Mitrakovic, and S.R. Schaps
National Institute of Standards and Technology
Boulder, Colorado

ABSTRACT

Stress has subtle effects on acoustic properties and exploiting these may lead to a field-usable system to detect and measure residual stress in the rims of railroad wheels. Acoustic birefringence is the underlying principle of operation. This is a measure of the relative difference in the propagation times of two shear waves polarized in radial and circumferential directions. The ultrasonic probe here is an EMAT (electromagnetic-acoustic transducer). This type of transducer requires little or no surface preparation and no acoustic couplant. The system operates in a pulse-echo mode. A short burst of shear horizontal waves travels through the rim thickness. The rotation of the EMAT determines the orientation of the polarization vector, radial or circumferential. Precise timing of echoes in both directions reveals the degree of birefringence. Changes are due to both stress state and metallurgical texture. Initial tests indicate it may be possible to separate these two. The Federal Railroad Administration now has this instrumentation for field tests.

MOST RAILROAD OPERATIONS in this country now deal with the transportation of freight in its many forms. Modern needs place ever greater demands on the elaborate networks that began developing early last century. Just one of the important elements in this complex system is the integrity of the many wheels that bear the loads. As massive as the cast-steel wheels used in the U.S. are, they might not seem to be a problem. Accident statistics, however, indicate otherwise. According to the Accident/Incident Bulletin of the Federal Railroad Administration (FRA) Office of Safety, broken wheels caused 134 accidents in the four year period of 1985-1988, and the cost was $27.5M. Besides the direct and indirect cost of such problems, there is the threat to human life and health due to the accidents themselves or the potential release of dangerous materials. Disruption to commerce at many levels is almost inevitable.

During their lifetime, wheels see a wide variety of sometimes extreme conditions. Among these are large static and dynamic mechanical loads, large-scale frictional wear, high temperatures from braking operations, environmental extremes, and exposure to potentially corrosive materials. All these factors operate to initiate tread cracks or aggravate any existing faults. According to the stress conditions (applied and residual), crack growth can be very slow and gradual or rapid, even explosive. The relative rarity of wheel failures under all these extreme conditions speaks well for the work done on their design and manufacture.

Wheel improvements will continue, but wear and tear are inevitable. For economic, environmental, and social reasons, replacement before failure is the preferred approach. Because of the costs and uncertainties involved, retirement-for-cause is favored, whenever possible, over replacement at predetermined intervals. Effective and efficient nondestructive evaluation can make it possible to evaluate fitness for service. For railroad wheels, as in most other applications, the first NDE test is visual examination by a skilled inspector. In practice, however, this simple approach has severe limitations. The stress state itself is not directly observable, although current regulations[1] look for evidence of a reddish-brown discoloration, indicative of a high-stress state. The ideal test method would be simple, reliable, unintrusive, inexpensive, and operator independent. No available techniques (X-ray, hole drilling, neutron diffraction) fulfill all these conditions perfectly, but ultrasonic testing offers a possible major step to an instrument for use in a railcar shop during maintenance operations or possibly at trackside.

ACOUSTIC BIREFRINGENCE AND STRESS

Birefringence (or double refraction) is the variation of the propagation velocity of a wave with the direction of the polarization vector. This property is probably most familiar as an optical phenomenon in some natural crystals, such as calcite. Some polymers exhibit a photoelastic effect; i.e., their birefringence is a function of mechanical stress. Such materials are available commercially and they permit a visual representation of the development of an applied surface stress field.

Acoustic waves in anisotropic materials exhibit birefringence. There are two principal causes for this: stress fields and metallurgical texture or preferred orientation of grains. While there are several ways to probe this anisotropy,[2,3] Fig. 1 illustrates the approach pursued in this work. We place a transducer on the front face of the wheel rim and launch an ultrasonic pulse through the thickness of the rim. This pulse will reflect and return to the transducer from the back face; its transit time is indicative of the velocity. The ultrasound is a horizontally polarized shear wave. There are two timing measurements necessary, with the polarization vector oriented along the radial (R) direction and tangential to the wheel circumference (θ). To achieve this, the transducer remains in the same location and we simply rotate it 90°. The distance traveled by the two pulses is exactly the same, so the experimental parameters are their respective arrival times.

We define the birefringence as the fractional change in the velocity of the two polarizations:

$$B = (V_\theta - V_R) / (V_\theta + V_R)/2 = (T_R - T_\theta) / (T_R + T_\theta)/2. \quad [1]$$

B is the birefringence; V is the velocity, and T is the transit time of waves polarized in the R and θ directions. (The travel distance is the same for both measurements.)

The functional relationship between birefringence and stress is:

$$B = B_0 + C_A(\sigma_\theta - \sigma_R).$$

B_0 is the birefringence due to the metallurgical texture. C_A is the stress acoustic constant. The σ's are the stresses in the principal directions.

The value of C_A for shear waves is quite small,[4] about -7.6×10^{-6} (MPa)$^{-1}$ or -5.2×10^{-5} (ksi)$^{-1}$ for steel wheels. Effectively, this means that the timing measurements must be very accurate (typically a few parts in 10^5, or <10 ns in a 100 μs interval). B_0 is usually a significant or major portion of the total birefringence, B. If we know B_0 (e.g., from a measurement on an unstressed specimen), the stress difference is then

$$\sigma_\theta - \sigma_R = (B - B_0) / C_A. \quad [2]$$

Usually, σ_R is negligible.

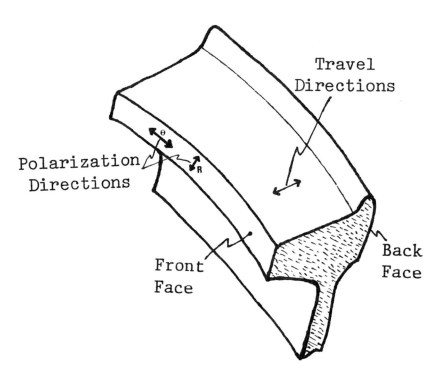

Figure 1. Schematic of a section of wheel rim indicating the travel and polarization directions of the ultrasonic pulse.

SYSTEM CONFIGURATION

Figure 2 shows a schematic of the basic units of the system. The EMAT (electromagnetic-acoustic transducer) is a device that generates and detects sound energy directly in an electrically conductive specimen.[5] A major advantage of this class of device is that they do not require the acoustic couplants usually associated with piezoelectric transducers. This simplifies operation and eliminates echoes and mode conversions that might occur at the interfaces the signal would otherwise have to traverse. Furthermore, they can operate effectively on rough and pitted surfaces.

The EMAT electronics are in two parts:
1. A high-current power amplifier that generates a gated pulse at 2 MHz. We use a 10-cycle pulse with a 78 Hz pulse-repetition-frequency. A trigger generated with each pulse starts the timer.
2. A low-noise, high-gain amplifier for the received signal.

The timing has three parts:
1. A digital gate circuit that locks onto a particular cycle of the echo signal and stops the timer.
2. A commercial time interval counter capable of nanosecond measurements and signal averaging.
3. An oscilloscope to observe the acoustic signal and the digital gate.

EXPERIMENTAL RESULTS

Early tests[6-9] compared measurements made with electromagnetic and more conventional piezoelectric transducers. The measurements made with the two systems were the same within experimental error. One important difference did emerge, however. To achieve good acoustic coupling between the PZT device and the wheel rim, it was necessary to mill a flat area on the rim, while the EMAT could operate over a rusted, pitted surface. This has obvious advantages for field use where the EMAT may require only modest wire-brushing to knock loose debris from the test surface.

The degree of birefringence is strongly dependent on the radial position of the transducer.[4] Figure 3 shows the convention used here. The zero position was at the approximate center of the width of the front rim face. Negative distances indicate positions closer to the axle, while positive distances mean sites nearer the tread. This position dependence is small, but, since it is comparable to the size of the effects of interest, it is essential that the position be carefully controlled. Some possible causes of this effect are signal interaction with the tread surface, the strong variation of the metallurgical texture, or a significant gradient in the stress state. These small shifts between measurements meant exerting considerable care in placing the transducers. To speed up this process, we constructed a wheel fixture to make operations easier and more reliable.

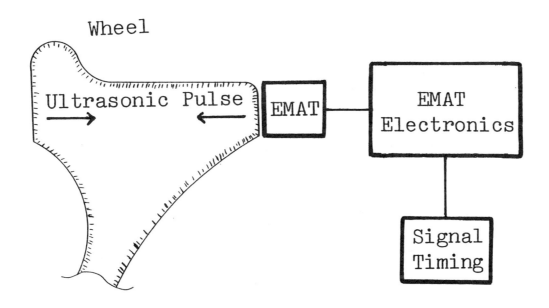

Figure 2. Block diagram of ultrasonic stress measurement system.

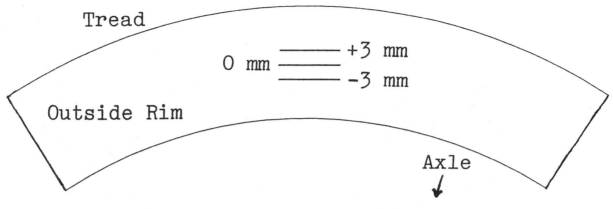

Radial Position Convention

Figure 3. This is the convention for radial position used in the work. The zero position is at about the mid-line of the front face. Negative positions are closer to the axle, while positive ones are closer to the tread.

Previous work[4] had already shown the sensitivity of acoustoelasticity to the stress state. However, the AAR Technical Center in Chicago lent us a pair of test wheels to demonstrate this effect in another way. While the wheel stress characterization was not sufficient to quantify or calibrate our measurements, the data did display the known and expected characteristics.

Both wheels had a saw cut along a radius, from tread to hub. At the outer (tread) end, there was a larger cut-out to accommodate the actuator of a small hydraulic press. With a hand-operated pump, it was possible to spread two sides of the wheel rim apart and introduce a hoop compressive load. For the measurements plotted in Fig. 4, the EMAT was positioned 180° from the actuator. After pumping the hydraulic pressure to pre-determined levels, the time measurements were made at several radial positions.

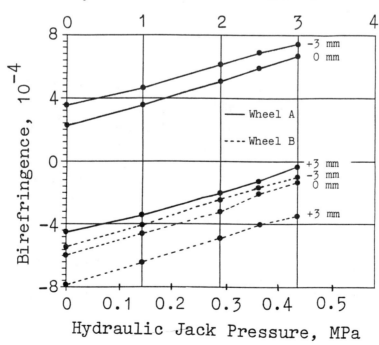

Figure 4. Birefringence in two AAR test wheels as hydraulic pressure is applied to generate compressive hoop stress. The ultrasonic tests were made at three radial positions on each wheel.

The plots in Fig. 4 reveal several characteristics:
1. The birefringence increases with higher pump pressure or greater compressive stress. Using the sign conventions of Eqs. [1] and [2], this confirms that C_A is negative.
2. If we assume that the rim hoop stress is linear with the pump pressure, Fig. 4 shows the stress is also linear with the birefringence, as per Eq. [2]. Note that all the data sets have the same slope or C_A.
3. The zero-pressure measurements from the two wheels are considerably different. This is most likely due to different residual stresses still remaining after the one radial saw cut. They may also have a different underlying metal texture.
4. As before, there is a strong radial dependence, with the most negative birefringence occurring nearest the tread.

Earlier reports of measurements on an intact wheel[6,9] indicated the presence of compressive residual stress in the rim. Subsequent saw cutting at the Transportation Test Center (TTC) in Pueblo, Colorado, gave qualitative, if not quantitative, confirmation when the wheel closed on the saw blade during the cutting.

As noted above, the metallurgical texture of the specimen contributes a significant fraction of the measured birefringence. This must be taken into account when calculating the stress (Eq. [2]). If the specimen is available before being stressed or there is always an unstrained reference specimen available that contains the same texture, the problem is simply to make two measurements and subtract. In practice, this is seldom possible - never for in-use tests of railroad wheels.

The wheels had a prominent texture produced by the casting process. Polishing and macroetching a thin slice from the rim of a wheel taken from service showed a large dendritic structure growing normally from each exterior surface. The dendrites are about 20-25 mm long as the result of grain growth during cool-down after casting. This large structural gradient indicates the importance of establishing a fixed EMAT position before making the timing measurements.

To help learn the magnitude of the texture problem as a prelude to working on a solution, TTC sent us two rim blocks each from ten different wheels. The manufacturer was the same for all wheels, but they were fabricated by four different plants. The rim blocks were about 20 cm long and nominally free of bulk residual stress. The measurements on these are in Fig. 5, grouped by production plant. Figure 6 combines the four data sets.

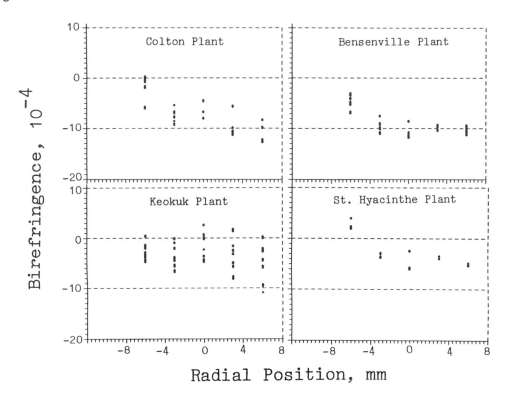

Figure 5. Birefringence measured on rim blocks from four production plants. The effect here is due solely to the metallurgical texture. There appear to be characteristic differences. An expanded data base is necessary to refine these trends.

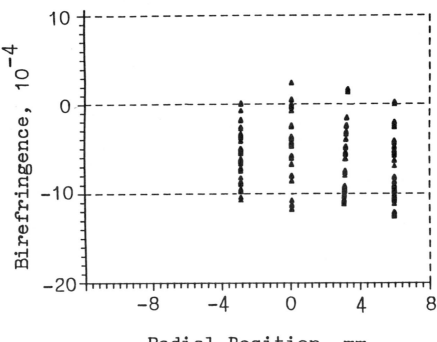

Figure 6. Collection of all the data in Figure 5. Even in this worst-case scenario, with no knowledge of wheel's origin, the uncertainty in the texture is still within acceptable limits.

Figure 6 gives a worst-case indication of the data scatter. A simple average of the values is $\simeq -5 \times 10^{-4}$ with a spread (δB_0) of $\pm 5 \times 10^{-4}$. This translates to a stress uncertainty ($\delta B_0/C_A$) of ± 65 MPa (10 ksi). Figure 5, however, indicates two possible refinements of this process:
1. Take into account the plant of origin (known from wheel ID markings).
2. Allow for the variation with radial position by curve fitting each plant's data (probably with a simple straight line).

With these procedures and a larger data base, it is possible to cut the texture uncertainty by a factor of almost two.

While considerably more data are necessary, these initial measurements are a positive indicator that this approach will have sufficient accuracy and sensitivity to become a useful tool.

ACKNOWLEDGMENTS

A considerable number of people have been involved in this work in various ways and at various time.

This work was funded by the Federal Railroad Administration, U.S. Department of Transportation under Donald Gray.

From the Association of American Railroads (AAR), we received a great deal of help and cooperation in obtaining test wheels. Of course, many tests would have been impossible without their collaboration. Our main contact has been Robert Larson, Jr. of the Transportation Test Center (TTC) in Pueblo, Colorado, and Dave Utrata of the Technical Center in Chicago. Some of the many others include Britto Rajkumar, Robert Vandeberg, and Robert Florom.

Both before and after his stay as a guest researcher in our laboratory, Professor Hidekazu Fukuoka of Osaka University was a valuable guide to the stress effects on ultrasound, particularly in railroad wheels.

Within our own lab, there have been many helpful ideas and considerable assistance at all stages. We are especially grateful to Dale Fitting, Todd McGuire, Jon Fowler, Peter Shull (now at the Dept. of Materials Science, Johns Hopkins University), Christopher Fortunko, Daniel Vigliotti, and still others. Tim Waldorf's skills were responsible for producing the wheel fixture. Avinoam Tomer (guest researcher from the Nuclear Research Center Negev, Beer Sheva, Israel) polished and etched the wheel rim cross section.

REFERENCES

1. Federal Railroad Administration, DOT, Code of Federal Regulations 49 CFR, ‡215.103 (h).
2. R.B. King and C.M. Fortunko, "Residual-Stress Measurements Using Shear-Horizontal Waves from Electromagnetic-Acoustic Transducers," NBSIR 84-3002 (March, 1984). (This is a collection of published papers.)
3. Y.-H. Pao, W. Sachse, and H. Fukuoka in "Physical Acoustics," Vol. XVII, Academic Press, New York (1984), 61-143.
4. H. Fukuoka, H. Toda, K. Hirakowa, and Y. Toya in "Wave Propagation in Inhomogeneous Media and Ultrasonic Nondestructive Evaluation," Vol. 6, G.C. Johnson, ed., ASME (1984), 185-193.
5. B.W. Maxfield and C.M. Fortunko, Materials Evaluation 41, 1399-1408 (1983).
6. A.V. Clark, H. Fukuoka, D.V. Mitrakovic, and J.C. Moulder in "Review of Progress in Quantitative Nondestructive Evaluation," Vol. 6B, D.O. Thompson and D.E. Chimenti, eds., Plenum Press, New York (1987), 1567-1575.
7. A.V. Clark, Jr., H. Fukuoka, D.V. Mitrakovic and J.C. Moulder, Ultrasonics 24, 281-288 (1986).
8. A.V. Clark, R.E. Schramm, H. Fukuoka, and D.V. Mitrakovic in "Proceedings: IEEE 1987 Ultrasonics Symposium," B.R. McAvoy, ed., Institute of Electrical and Electronic Engineers, New York (1988), 1079-1082.
9. A.V. Clark, H. Fukuoka, D.V. Mitrakovic, and J.C. Moulder, Materials Evaluation 47, 835-841 (1989).

Practical Applications of Residual Stress Technology, Conference Proceedings, Indianapolis, Indiana, USA, 15-17 May 1991

Residual Stresses, Great and Small

G.T. Blake
Wiss, Janney, Elstner Associates, Inc.
Northbrook, Illinois

ABSTRACT

The hole-drilling strain-gage method for determining residual stresses has been used for many years. Many technical papers have been written describing the method, the accuracy, effectiveness, precision, and suggesting ways to enhance and improve the use of the technique. Very few papers have been presented demonstrating practical applications. This paper describes three dissimilar projects where the method was successfully used to solve either a technical problem or provide a useful service.

AS A REVIEW, the hole-drilling strain-gage method involves adhesively bonding a special rectangular (0°, 90°, 135°) strain gage rosette to the surface of an elastic material, drilling a small hole at the intersection of the center lines of the three strain gages and measuring the relieved strains. The principal residual stresses and their orientation are calculated using the measured strains in a series of equations.[1] The method assumes that the stresses are elastic, uniform or linear with depth. New techniques which have been recently developed to measure nonlinear residual stresses, are not discussed.

The following case histories are not intended to give you an analysis of the results of the tests, and I have purposely omitted equations and data. Rather, I hope these projects will stimulate you into thinking about your own applications where this method may be used successfully.

MEASUREMENT OF TRIAXIAL STRESSES IN RAILROAD RAILS[2]

EXPERIMENTAL WORK - Cracked railroad rails are a constant maintenance problem. It has been suspected that some cracking is due to accumulated residual stresses. Under the sponsorship of the Track/Train Dynamics program of the Association of American Railroads, a project was developed to study this problem. This project included development of a method for measuring triaxial residual stresses in railroad rails. The method applies specifically to residual stresses that are constant along the rail except near cut ends. This means that one of the three principal stresses is in the longitudinal direction and that there are no shear forces on a transverse plane through the rail.

The method involved three measurement and analysis steps. First, the method of sectioning was used to determine a set of longitudinal (z) residual stresses neglecting the Poisson effect of the transverse residual stresses. Second, a set of transverse (x and y) residual stresses that also neglected the Poisson effect of the longitudinal stresses was measured on a transverse slice of the rail using the hole-drilling strain-gage method. Third, an iterative mathematical procedure involving a finite-element model of the slice was used to correct for the Poisson effects of the longitudinal and transverse stresses on each other and determine the three dimensional stress field.

SPECIMENS - Samples of six rails were used for this study. Rails No. 1 and 2 had never been placed in service. Rails No. 3, 4, 5, and 6 were taken from service after carrying cumulative loads of approximately 150, 200, 450, and 600 million gross tons, respectively. Specimens from each rail sample were prepared as shown in Fig. 1.

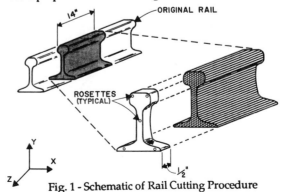

Fig. 1 - Schematic of Rail Cutting Procedure

LONGITUDINAL STRAINS - The method of sectioning was used to determine the longitudinal residual strains, which correspond to the longitudinal stresses if the Poisson effect is neglected. Briefly, this method consists of measuring the relaxation of strain that occurs when a short length is cut from a member and this piece is further cut into small longitudinal bars or elements. The longitudinal residual strain that existed in the bar before any cuts were made is equal to the measured relaxation in strain and is opposite in sign to this measured strain.

This method was applied as follows. First, a 356 mm (14 in.) length was cut from the rail, measuring the change in length due to the cut. Next, the two ends of the 14 in. length were machined to provide near-parallel transverse surfaces. A 7.6 mm (0.3 in.) grid of vertical and horizontal lines was marked on one cut surface and the piece of rail was placed in a special fixture, as shown in Fig. 2.

Fig. 3 - Zero longitudinal strain measurements

A longitudinal saw cut was then made along each grid line. Each cut started at one end of the rail and continued a distance of 254 mm (10 in.) into the rail. Thus, the rail was cut into 7.6 mm (0.3 in.) square bars that were 254 mm (10 in.) long and were connected to a solid piece of rail 102 mm (4-in.) long (Fig. 4).

Fig. 2 - Special fixture for longitudinal strain measurements

The length of the piece at the center of each grid square was measured by a dial-gage micrometer with a least reading of 0.0025 mm (0.001 in.), as shown in Fig. 3. The plastic template shown was used to locate the measurement points. Since the end cuts were not perfectly flat, the original length measurements for the various grid squares differed slightly.

Fig. 4 - Saw-cutting longitudinal bars

Some of the bars, especially those around the periphery, bowed significantly as a result of cutting. To eliminate the bowing, shims were placed between the bars and wire was wound tightly around the outside of the rail. The thickness of the shims was equal to the thickness of the saw cuts, which was about 1.6 mm (1/16 in.). Finally, the relaxed length of each bar was measured in the special fixture. The difference between the initial reading before longitudinal cutting and the final reading for each bar indicates the total deformation of the bar as a result of the longitudinal cuts.

POISSON CORRECTION - A finite-element analysis of a 12.7 mm (1/2 in.) thick transverse slice was used to calculate the transverse stresses, and the shear stress caused by the longitudinal stresses. The general-purpose program ANSYS was used for the analysis. The slice was modeled with two layers of three-dimensional isoparametric solid elements. The longitudinal stresses were applied to both surfaces of the modeled slice. A constant stress (or pressure) was applied over each element, but the stresses applied to various elements were different.

TRANSVERSE STRESSES - A 13 mm (1/2 in.) thick transverse slice was cut from the rail adjacent to the portion removed for sectioning, and was used to measure the transverse biaxial stresses in the x and y directions neglecting the Poisson effect. Each slice was chemically polished to remove any metal disturbed by the slicing operation and eliminate any mechanically induced residual stresses in the cut face. The hole-drilling method was used to measure the stresses at approximately 28 locations as shown in Fig. 5.

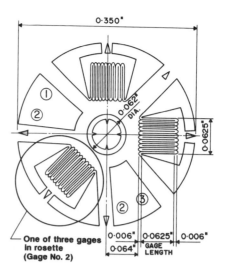

Fig. 6 - Special rosette for use in hole-drilling method

An S.S. White industrial Airbrasive Unit Model F, shown in Fig. 7, was used as the hole-forming unit in this study. The jet head was leveled, aligned, and held in place over the strain-gaged workpiece using a Vishay RS-200 milling guide.

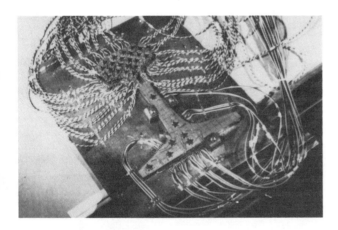

Fig. 5 - Hole-drilling strain gage locations for transverse specimen

Micro-Measurements Type EA-06-062RE-120 strain-gage rosettes were used, as shown in Fig. 6. These gages, which are specifically designed for residual strain measurements, were all from the same lot with a nominal gage factor of 1.99. The gages were self-temperature compensating when bonded to steel. Each three-element rosette had a mean gage circle diameter of 0.202 in. The gages were bonded to the rail section using M-Bond 200 adhesives.

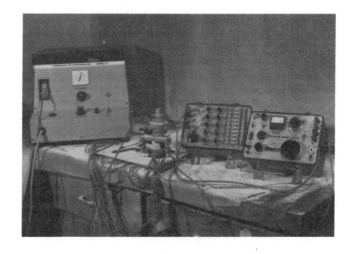

Fig. 7 - Air-Brasive drill and data acquisition equipment

The gage area surrounding the hole location was protected by a layer of masking tape to prevent abrasion by stray particles. Previous work by Beaney and Proctor[3] indicated that the residual strains around a drilled hole are completely relieved when the hole depth equals the diameter of the drilled hole. Therefore, the holes in this study were abraded to a depth slightly greater than one diameter about 1.65 mm (0.065 in.) diameter). The holes were formed using 50-micron aluminum-oxide abrasive powder at 590-kPa (85-psi) air pressure and a vibrator setting of five. After the hole was formed, the final diameter was measured microscopically and recorded.

ANALYSIS - All of the measured and calculated strains were then used in an iterative FEM computer program to where all forces and moments were balanced to produce the three dimensional stress field.

CONCLUSIONS - The method described herein is satisfactory for determining the triaxial residual stresses in rails if these stresses are constant along the length except near the ends. The investigation of the six rails showed that the Poisson effect of the longitudinal stresses on the transverse stresses is small and can be neglected in most applications.

MEASUREMENTS OF RESIDUAL STRESSES IN WIND TUNNEL WELDMENTS

A 6.1 x 15.2 m (20 ft x 50) ft wind tunnel built during World War II was being considered for upgrading from low altitude use to a high altitude chamber. A plan view of the tunnel is shown in Fig. 8.

Fig. 8 - 50 ft. diameter section of wind tunnel

The welded circular sections (up to 15.5m (51 ft) in diameter) were fabricated by unskilled welders using whatever materials were available during war time.

Vacuum testing during a previous renovation indicated that the section of the tunnel earmarked for conversion leaked badly and it was suspected that detrimental residual stresses existed at the welds.

As a part of the work required to estimate the costs of renovation, it was necessary to make a residual stress analysis of the structure to determine what structural alterations should be considered.

EXPERIMENTAL WORK - A total of 60 measurements were made on the interior and exterior of the wind tunnel. Inside and outside measurement locations were mirror images of each other so that the state of stress of both surfaces could be measured at the same location.

Micro-Measurements Type EA-06-062RE-120 strain gages were also used in this study. A Measurements Group Model RS-200 Drilling Guide and Hi-Speed Accessory powered by bottled nitrogen was used to drill the strain relieving holes in ten equal increments. A 1.57 mm (0.062) in. diameter carbide cutter drilled each hole to a depth of about 1.91 mm (0.075 in.). A calibrated Measurements Group Model P350 Strain Indicator and Model SB-1 Switch, Balance Instrument were used to measure the strain-gage output at each depth increment.

The calculations of the principle stresses were made using the equations in ASTM E837.[1] Young's Modulus of 20×10^{10} Pa (29×10^6 psi) and Poisson's Ratio of 0.3 were used in the calculations.

Many of the measurement locations in the tunnel were located on the side walls. In order to provide an immovable base for the drilling guide, the feet of the guide had to be bonded to the walls at each location. To do this, a template of the feet locations was used to locate their position in relation to the bonded strain gage as shown in Fig. 9. Fast curing Hydra-Cal dental cement was used to bond the feet. Hydra-Cal was preferred to epoxy because it is easy to remove in preparation for the next set up.

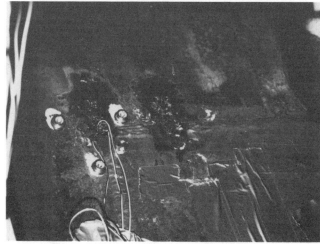

Fig. 9 - Adhesively bonded milling guide feet

Fig. 10 - Typical bonded milling guide set-up

Fig. 10 shows the technique being used in the upside down position. Some external measurement locations were near corners and required special care to ensure that the axis of the drilling guide was installed radial to the surface being measured. A typical corner location is shown in Fig. 11.

Results of these tests indicated the presence of stresses in the structure ranging from near zero to well beyond yield.

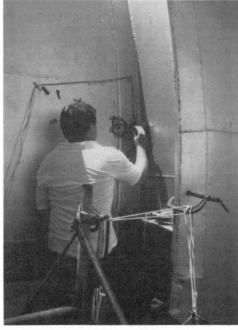

Fig. 11 - Measurement at an exterior corner weld

RESIDUAL STRESS MEASUREMENTS IN U-BENDS OF 5/8 IN. (16 MM) PERIMETER STAINLESS STEEL TUBING[4]

Our third case history involves measuring residual stress in material that has been mechanically deformed and then heat-treated to relieve the residual stresses caused by the deformation.

Stainless steel tubing is used primarily by electric power, chemical and petrochemical industries in heat exchangers. In order to achieve the most efficient heat exchanger design, lengths of tubing are bent into U-bend shapes to fit inside the equipment housings. U-bends are formed by clamping one end of each tube to be bent and mechanically swinging the free end against a circular reaction. The mechanical work in bending the tube causes plastic deformation and induces residual stresses throughout the length of the bend. Since residual stresses in the tubes removed by heat-treating, the electric-resistance heating of the bent portion of the tube to about 1121°C (2050° F) and then allowing the tube to cool using forced air.

Recent tube U-bend failures in power industry heat exchangers have caused concern that heat-treating is not sufficiently removing the residual stresses in the tubes. In order to improve the quality control of the heat-treating process and prevent stress-related failures of their products, the Plymouth Tube Company retained Wiss, Janney, Elstner Associates, Inc. (WJE) to measure the residual stresses in specimens taken from production runs of tubing.

EXPERIMENTAL WORK

<u>Specimens</u> - Typical test specimens of 64 mm (2-1/2 in.) and 89 mm (3-1/2 in.) diameter U-bends, 203 mm (8 in.) long have wall thicknesses of 0.9 mm (0.035 in.) and 2.3 mm (0.090 in.), respectively. A 64 mm (2-1/2 in.) diameter U-bend specimen is shown in Fig. 12.

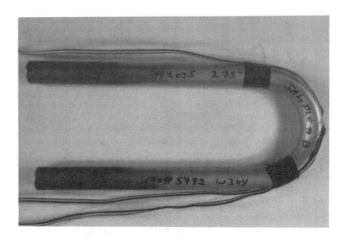

Fig. 12 - Typical U-bend specimen

Specimen Calibration - The Blind-Hole Strain Gage Method equations contained in ASTM E837-85[1] that are used to calculate surface residual stresses, are based on behavior which occurs when a hole is drilled into a thin, wide plate. The resulting hole has a circular cross section, straight sides and a flat bottom perpendicular to the sides. The equations are also based on measuring stress at a point and do not account for the averaging effect of the strain gages which have a finite surface area under the strain-gage grid. In addition, the equations assume a constant stress through the plate thickness. Because of the geometric differences between the surfaces of a tube and a thin, wide plate, the strain relief behavior of drilled tube specimens is assumed to be be different. Therefore, it was necessary to account for those differences by conducting calibration tests on tube samples.

Calibration Specimens - Several different calibration techniques have evolved while trying to refine the material coefficients to truly represent the behavior of the bent tubes. One such technique uses sections of straight tubing in a variation of the calibration procedure suggested by ASTM.[1]

Four 152 mm (6 in.) long calibration specimens were made from 15.9 mm (0.625 in.) diameter Type 304 stainless steel tubing. One pair of tube specimens had nominal wall thicknesses of 0.9 mm (0.035 in.) The other pair had nominal wall thicknesses of 2.4 mm (0.095 in.). Machined steel plugs, 25 mm (1.0 in.); long were machined to slip fit into the ends of each specimen to prevent crushing during gripping.

Four Micro Measurements Type CEA-06-125 UW-125 single element strain gages, located at 90° to each other, were bonded to the outside surface along the longitudinal axis of each specimen at midlength. One Micro Measurement Type EA-06-031RE-120 strain gage rosette was bonded at the center of each 0.035 in. (0.9 mm) wall specimen. A type EA-06-062RE-120 strain gage strain gage rosette was bonded to the center of each 2.4 mm (0.090 in. wall specimen. Gage No. 3 of each rosette was located along the longitudinal axis of the tubes. All gages were bonded using M-Bond 200 adhesives and normal strain gage bonding practice. One of the gaged calibration specimens is shown in Fig. 13.

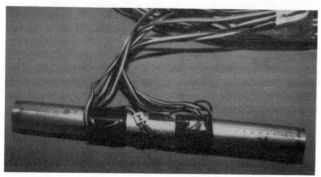

Fig. 13 - Typical calibration specimen

Specimen Fixtures - Collet blocks 44 mm (1.5 in.) square were used to grip the specimens. One end of each collet block was adapted to anchor a 152 mm (6 in.) length of 13 mm)0.5 in.) diameter steel chain. The other ends of the chain were linked to 49 mm (2 in.) diameter eye bolts used for testing machine connections, as shown in Fig. 14.

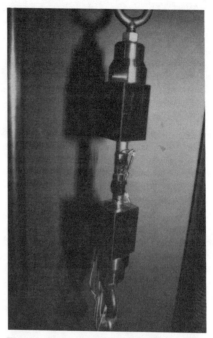

Fig. 14 - Typical tensile calibration set-up

Test Procedure - Each specimen was placed in a universal testing machine such that the tube specimen was loaded axially. The strain gages were connected to a Vishay P3500 Strain Indicator through an SB-1 Switch/Balance instrument. All gages were balanced to indicate zero strain at zero load. Tensile loads up to about 33 percent of the proportional limit of the material were applied in two equal increments. This was about 3827N (860 lbs) and 9612N (2160 lbs) for the 0.9 mm (0.035 in.) and 2.9 mm (0.095 in) wall specimens, respectively. Strain-gage readings were taken at each load increment. After reducing the load to zero, each tube calibration specimen was removed from the testing machine and using the Measurements Group RS-200 Milling Guide and Hi-speed turbine accessory, a hole depth increment of 0.05 mm (0.002 in.) was drilled at the rosette location. The strain gages were then rezeroed and the specimen was placed in the testing machine and reloaded as before. This procedure was repeated in 0.05 mm (0.002 in.) increments until the hole depth was 0.25 mm (0.010 in.). The depth increments were then increased to 0.13 mm (0.005 in.). The final hole depth for the 0.09 mm (0.035 in.) and 2.9 mm (0.095 in.) wall specimens was 0.9 mm (0.035 in.) and 2.3 mm (0.090 in.), respectively.

COMPUTATION OF CALIBRATION COEFFICIENTS - Standard calibration coefficients used in ASTM 837 have only been established for the calculations of residual stresses in thin flat plates. Values for determining residual stresses in tube specimens were calculated from the incremental strain-gage rosette readings observed before drilling, and after drilling at the two calibration loads. The calibration stresses used in these tests were determined from the four axial strain gages using Hooke's Law. These new values are now used in a computer program developed for making routine measurements of Stainless Steel U-bends.

In practice, Plymouth Tube Company cuts heat-treated U-bend specimens from production runs of tubing. The specimens are carefully packed and shipped to WJE for testing. Micro-Measurements type EA-06-062RE-120-option SE strain-gage rosettes with a gage-circle diameter of 5.1 mm (0.202) in. are used to measure the relieved strains in the 2.3 mm (0.090 in.) wall specimens and type EA-06 031RE-120-option SE strain gages are used for the 0.9 mm (0.035 in.) wall specimens.

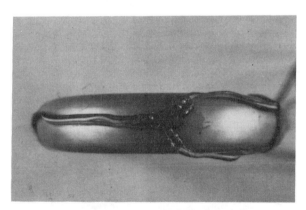

Fig. 15 - Typical tested U-bend specimen

After completing the test, which takes about 2 hours for each specimen, the computerized results are mailed to Plymouth Tube for inclusion with the shipping invoice and other material reports. Current practice limits the maximum principal residual stress to (345 x 10^6 Pa (5000 psi). If the results of the measurements are outside the limit, Plymouth Tube makes adjustments to its heat treating procedures and submits additional specimens for retesting.

SUMMARY - The purpose of this paper has been to present three practical applications of using the hole-drilling strain gage method to determine residual stresses. We have shown how the technique can be used in conjunction with other techniques to determine residual stresses in the laboratory. We have demonstrated the techniques that can be used in field applications and we have shown how the technique can be used as a quality assurance tool.

For those applications where measurements of surface residual stresses can be made using a semidestructive method, the hole drilling strain gage method is a relatively inexpensive, dependable technique.

References

1. "Determining Residual Stresses by the Hole-Drilling Strain-Gage Method," American Society for Testing on Materials Standard Test Method E-837-85

2. C. G. Schilling and G. T. Blake, "The Measurement of Triaxial Residual Stresses in Railroad Rails - Measurement and Analysis Techniques," Experimental Techniques, Vol. 8, No. 9, (September, 1984)

3. E. M. Beaney and E. Proctor, "A Critical Evaluation of the Centre Hole Technique for the Measurement of Residual Stresses," Rep. RD/B/N2492, Berkeley Nuclear Laboratories (November, 1972)

4. G. T. Blake, "Residual Stress Measurements in U-Bends of 5/8 Inch (16 mm) Diameter Stainless Steel Tubing Using the Blind-Hole Strain Gage Method," American Society of Mechanical Engineers, 87-JPGC-PWR-25 (1987)

Practical Applications of Residual Stress Technology, Conference Proceedings, Indianapolis, Indiana, USA, 15-17 May 1991

The Use of X-Ray Diffraction to Determine the Triaxial Stress State in Cylindrical Specimens

P.S. Prevey and P.W. Mason
Lambda Research, Inc.
Cincinnati, Ohio

ABSTRACT

A method of determining the axial, circumferential and radial residual stress distributions in cylindrical specimens is described. The axial and circumferential residual stresses are measured directly by x-ray diffraction at the free cylindrical surface exposed by machining and electropolishing. The radial stress component is then calculated from an integral of the circumferential stress at the free surface as a function of depth by the method of Moore and Evans.

The method is applicable only to cylindrical samples with rotationally symmetrical stress distributions from which complete cylindrical shells are removed for subsurface measurement. The method does not require prior knowledge of the stress-free lattice spacing, and thus provides a means of verifying neutron and x-ray diffraction methods of full tensor stress determination. The stress-free lattice constant, d_o, is also calculated as a function of depth from the sum of the principal stresses.

Application of the method, to determine the triaxial residual stress distribution in an induction hardened 1045 steel multi-axial fatigue specimen, is described. The variation in the stress-free lattice spacing of the (211) planes with depth is estimated through the hardened case and into the core material.

INTRODUCTION

The classic $Sin^2\psi$ method of x-ray diffraction residual stress measurement (and the single-exposure and two-angle techniques derived from it) [1] is based upon a model of plane stress at the free surface of the sample. No stress normal to the surface is assumed to exist in the thin layer (on the order of 10 μm) effectively penetrated by the x-ray beam. Subsurface residual stress distributions are measured by removing layers of material in a manner which does not induce stresses. The stress distributions in directions lying in the plane tangent to the surface can be determined as a function of depth, provided stress relaxation caused by layer removal is trivial or can be calculated. Corrections for stress relaxation have been developed by Moore and Evans [2] for simple geometries and symmetrical stress fields.

The stress in the direction normal to the surface cannot generally be determined by x-ray diffraction using the plane stress model. Both x-ray and neutron diffraction techniques are available to determine the full stress tensor. However, the full stress tensor cannot be calculated without prior knowledge of the unstressed lattice spacing, d_o. For many practical samples, the unstressed lattice spacing may be difficult to determine or composition-dependent, and vary with depth into the sample surface, as in carburized, nitrided or induction hardened steels.

In the singular case of cylindrical samples with rotationally symmetrical stresses, it is possible to determine the axial, circumferential and radial components of residual stress as functions of depth by applying the method of Moore and Evans, provided complete cylindrical shells are removed to expose each subsurface layer. The radial stress component is calculated as an integral of the circumferential stress measured on each exposed cylindrical surface as a function of depth.

As each cylindrical surface is exposed, the sum of the principal stresses can be determined and the value of the stress-free lattice spacing, d_o, calculated (assuming plane stress at the exposed free surface) if the x-ray elastic constants are known. The variation in the stress-free lattice spacing with depth due to the carbon gradient and the percent martensite formed during heat treatment can then be determined.

The method is of interest, not only for the determination of the three-dimensional stress distributions in cylindrical, rotationally symmetrical parts, but also as a means of independently determining the three-dimensional stress state and the stress-free lattice spacing distribution for comparison to neutron and x-ray diffraction full-tensor stress measurement methods.

THEORY

Assuming a cylindrical sample (either a solid rod or a tube of inside radius R_1) with rotationally symmetrical stresses, Moore and Evans [2] developed closed-form solutions for the true radial, axial and circumferential residual stress distributions at any radius r_1 calculated from only the circumferential and axial stress distributions measured as functions of depth by removing cylindrical shells:

$$\text{EQ 1} \quad \sigma_r(r_1) = -\left(1 - \frac{R_1^2}{r_1^2}\right)\int_{r_1}^{R}\left(\frac{r^2}{r^2 - R_1^2}\right)\frac{\sigma_{\theta m}(r)}{r}\,dr$$

$$\text{EQ 2} \quad \sigma_z(r_1) = \sigma_{z m}(r_1) - 2\int_{r_1}^{R}\left(\frac{r^2}{r^2 - R_1^2}\right)\frac{\sigma_{z m}(r)}{r}\,dr$$

$$\text{EQ 3} \quad \sigma_\theta(r_1) = \sigma_{\theta m}(r_1) + \left(\frac{r_1^2 + R_1^2}{r_1^2 - R_1^2}\right)\sigma_r(r_1)$$

For the case of a solid bar, as in this application, R_1 equals zero. The radial stress component is calculated from the measured circumferential stress using Equation 1. The axial and circumferential stress distributions are corrected for stress relaxation caused by layer removal, using Equations 2 and 3.

Assuming that a condition of plane stress exists on the electropolished surface, free of any machining or grinding deformation, the lattice spacing will depend upon the stresses present in that surface as,

$$\text{EQ 4} \quad d(\psi) = \left(\frac{1+\nu}{E}\right)\sigma_\phi d_o \sin^2\psi - \left(\frac{\nu}{E}\right)(\sigma_1 + \sigma_2)d_o + d_o$$

where ψ is the angle of tilt from the surface normal, σ_1 and σ_2 are the principal stresses, ν and E are Poisson's ratio and Young's Modulus, respectively, and d_o is the stress-free lattice spacing. Equation 4 is the basis for the $\sin^2\psi$ method of residual stress determination by x-ray diffraction.

In the direction parallel to the surface normal, $\psi = 0$; therefore the observed lattice spacing depends only upon the sum of the principal stresses. Because the sum of the principal stresses equals the sum of any two perpendicular stresses, the sum of the circumferential and axial components may be substituted, and the unstressed lattice spacing is given by,

$$\text{EQ. 5} \quad d_o = \frac{d(\psi = 0)}{1 - \left(\frac{\nu}{E}\right)(\sigma_A + \sigma_C)}$$

where σ_A and σ_C are the axial and circumferential residual stresses measured on the free surface exposed by electropolishing. The x-ray elastic constants in the (hkl) direction of interest would generally be measured directly. [3] The value of d_o could be further corrected for systematic instrumental error, which was not undertaken in this investigation.

SAMPLE PREPARATION

The sample used in this study was prepared as part of the Society of Automotive Engineers Fatigue Design and Evaluation Committee's study of multi-axial fatigue life prediction. The sample was reportedly manufactured from induction hardened, hot rolled, 1045 steel, and was identified as sample 49X65601B. The sample had been fatigued to one-half of its anticipated cyclic life of 750,000 cycles in 5000 N·M multi-axial fatigue. The dimensions of the sample are shown in Figure 1.

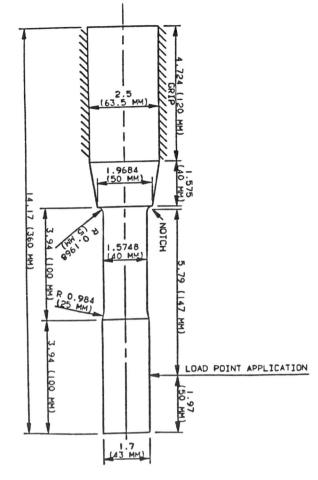

Fig. 1 SAE 1045 Steel Induction Hardened Multi-axial Fatigue Sample

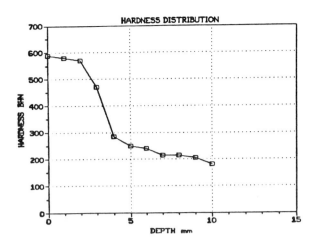

Fig. 2 Brinnell Hardness Distribution Through the Induction Hardened Case

The hardness distribution developed by induction hardening was measured on similar samples[4] and is depicted in Figure 2, indicating a hardened case depth on the order of nominally 4.5mm.

TECHNIQUE

Prior to x-ray diffraction measurement, the large grip end of the sample and the 100mm extension were removed from either end of the cylindrical gage section to facilitate handling during machining and measurement. Stress relaxation, caused by sectioning to reduce the sample length, was assumed to be negligible.

X-ray diffraction residual stress measurements were made in the longitudinal and circumferential directions on the uniform cylindrical gage section at a location 10mm from the point of tangency with the notch radius. Uniform cylindrical shells were removed from the gage section by first turning with a carbide tipped cutting tool, and then electropolishing to remove a minimum of 0.25mm of material over a local area approximately 1.5cm square to eliminate any residual stresses produced by turning.

X-ray diffraction residual stress measurements were performed using a two-angle technique employing the diffraction of chromium K-alpha radiation from the (211) planes of the BCC structure of the 1045 steel. The diffraction peak angular positions at each of the ψ tilts employed for measurement were determined from the position of the K-alpha 1 diffraction peak separated from the superimposed K-alpha doublet assuming a Pearson VII function diffraction peak profile in the high back-reflection region.[5] The diffracted intensity, peak breadth and position of the K-alpha 1 diffraction peak were determined by fitting the Pearson VII function peak profile by least squares regression after correction for the Lorentz polarization and absorption effects, and for a linearly sloping background intensity.

Measurements were performed on a computer-controlled Huber diffractometer, instrumented with a scintillation detector in a Bragg-Brentano geometry. The irradiated area was nominally 4mm x 4mm. The value of the x-ray elastic constant, $E/(1+\nu)$, required to calculate the macroscopic residual stress from the strain measured normal to the (211) planes of 1045 steel, was not determined during the course of this investigation. The data were reduced using constants previously determined for 1050 steel.[3]

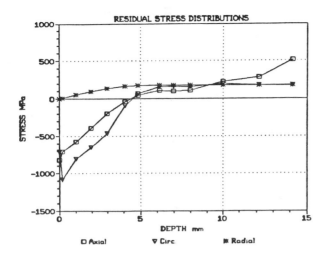

Fig. 3 Triaxial Stress Distributions at a Point 10mm from the Notch in the Cylindrical Gage Section

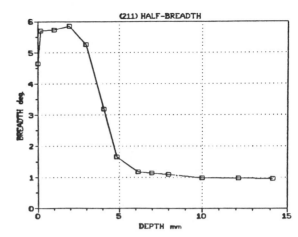

Fig. 4 (211) Diffraction Peak Width Distribution

All data obtained as a function of depth were corrected for the effects of the penetration of the radiation employed for residual stress measurement into the subsurface stress gradient.[6] Systematic errors caused by instrument misalignment and sample displacement were monitored per ASTM E915, and found to be less than +-14 MPa.

RESULTS AND DISCUSSION

The axial, circumferential and radial residual stress distributions are shown in Figure 3. The axial and circumferential results have been corrected using Equations 2 and 3, and the radial results are calculated from Equation 1.

The axial and circumferential stresses rise from maximum compression near the surface to cross into tension at a depth of nominally 4.5mm. The axial stress reaches nominally 500 MPa at a depth of 14mm. The radial stress component is necessarily zero at the surface, and rises through the compressive case to reach a nearly constant value on the order of 180 MPa at a depth of approximately 5mm.

The width of the (211) K-alpha 1 diffraction peak, separated from the K-alpha doublet by Pearson VII peak profile fitting,[5] is shown in Figure 4, without correction for instrumental broadening. The results show a sharp reduction in peak width at the surface, possibly the result of surface decarburization. In the hardened case, the diffraction peak width is on the order of 5.8 deg. to a depth of nominally 2.5mm. The peak width then diminishes rapidly through the remaining portion of the case to approach a uniform peak width of nominally 1.0 deg. in the soft core material. Comparison with the mechanically measured Brinnell hardness shown in Figure 2 indicates a similar trend. The sharp reduction in peak width at the surface is attributed to the shallow penetration of the x-ray beam, revealing a thin decarburized surface layer.

The subsurface distribution of the stress-free lattice spacing of the (211) planes, calculated from Equation 5, is shown in Figure 5. The results show a reduced lattice spacing near the surface, attributed to possible decarburization, followed by an increase in lattice spacing to nominally 1.1709 Å. The stress-free lattice spacing then diminishes with increasing depth through the case and reaches a value on the order of 1.1702 Å in the softer core material at a depth of

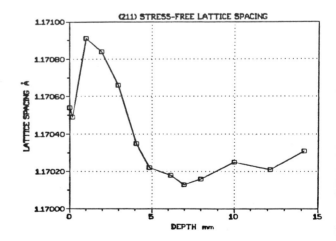

Fig. 5 (211) Stress-Free Lattice Spacing Distribution

nominally 5mm. The variation in the lattice spacing within the core and the reduced lattice spacing immediately beneath the surface are not completely understood. The results presented are the average of three repeat measurements of the lattice spacing at $\psi = 0$, each with the sample repositioned. The experimental error estimated from the repeat measurement is $\pm 3 \times 10^{-5}$ Å. The variation observed in the core between depths of 5mm and 14mm appears to exceed the estimated experimental error.

CONCLUSIONS

The method of correcting x-ray diffraction data for stress relaxation resulting from layer removal developed by Moore and Evans has been applied to determine the triaxial stress state throughout most of the volume of an induction hardened 1045 steel axle. The triaxial stress distributions obtained appear to correlate with the observed hardness variation, and provide for static equilibrium in the body.

A simple method of calculating the stress-free lattice spacing as a function of depth has been demonstrated. The results reveal a significant variation in the unstressed lattice spacing, d_o, through the induction hardened case and into the core of the material.

The method described, generally applicable to any cylindrical specimen with rotationally symmetrical stresses, appears to provide a novel method of determining both triaxial stress and the stress-free lattice spacing distributions with depth for comparison to neutron and x-ray diffraction solutions of the full stress tensor.

ACKNOWLEDGEMENTS

The authors gratefully acknowledge the assistance of Mr. Thomas Cordes of the John Deere Corporation, and Dr. Peter Kurath of the University of Illinois, for providing the test sample, and Dr. Henry Prask of the NIST for discussion of the calculation of the stress-free lattice spacing.

REFERENCES:

(1) M. E. Hilley, ed., Residual Stress Measurement by X-ray Diffraction, SAE J784a, SAE, Warrendale, PA 1971, p. 62.

(2) M. G. Moore and W. P. Evans, Trans. SAE, Vol. 66, 1958, p. 340.

(3) P. S. Prevey, ADV. IN X-RAY ANALYSIS, Vol. 20, 1977, p. 345.

(4) J. K. Oshsner, SAE Fatigue Design and Evaluation Committee Correspondence, June 1, 1988.

(5) P. S. Prevey, ADV. IN X-RAY ANALYSIS, Vol. 29, 1986, p. 103.

(6) D. P. Koistinen and R. E. Marburger, Trans. ASM, Vol. 51, 1959, p. 537.

Neutron Diffraction Measurements of Residual Stress Near a Pin Hole in a Solid-Fuel Booster Rocket Casing

J.H. Root, R.R. Hosbons, and T.M. Holden
Atomic Energy of Canada Ltd. Research
Chalk River, Ontario, Canada

ABSTRACT

Elastic strains in the axial, hoop and normal directions have been measured by neutron diffraction at about sixty locations between the pin hole and the end of the tang of a solid-fuel booster rocket casing. The largest strains were found to be near the pin hole in the region that yields under a combination of pin bearing and hoop tension loadings. Here, the maximum elastic compressive hoop strain was $(-23 \pm 2) \times 10^{-4}$ and the tensile axial and normal strains were $(12 \pm 2) \times 10^{-4}$. The detailed strain profiles were compared with the predictions of a finite-element calculation based on a classical incremental elastic-plastic material model. The calculations were found to be in qualitative agreement with the neutron diffraction results. This work was done in association with Thiokol Corp.

THE CASES OF THE MOTORS for NASA's Space Shuttle booster rockets are assembled from a number of segments, each machined from a separate forging. The forgings are made from D6AC, a high strength, low alloy steel. This material is known to be moderately susceptible to stress-corrosion cracking if exposed to sufficiently high tensile stresses for a prolonged period of time in a water or humid air environment. During proof testing of the case segments, local regions yield. When the proof test pressure is released, local elastic residual stress fields must develop to maintain compatibility between the yielded and non-yielded regions of the case. It is important that design engineers be able to determine the residual stresses arising from local yielding so they can identify potential increases in the susceptibility to stress-corrosion cracking.

NEUTRON DIFFRACTION

The physical principles involved in measuring internal strain and stress with neutron diffraction[1] are similar to those involved in the measurement of strain by the more conventional technique of X-ray diffraction.[2] Neutrons are diffracted by the planes of atoms in crystalline materials. Peaks in the diffracted intensity are observed when the Bragg condition is satisfied. This condition is given by the expression

$$\lambda = 2d \sin \theta \qquad (1)$$

where λ is the neutron wavelength, d is the interplanar spacing and 2θ is the angle between the incident and the diffracted beams. In a state of tension the interplanar spacings are expanded. The strain, ε, is a dimensionless quantity, the fractional change of d from that in a stress-free reference sample, d_0,

$$\varepsilon = (d/d_0 - 1). \qquad (2)$$

Thus, careful determinations of 2θ and λ give direct measurements of d and the strain. In a diffraction experiment, the strain component determined is the one directed along the bisector of the incident and diffracted beams. A number of strain components are determined by orienting the sample so that different sample directions lie along this bisector.

The primary advantage of neutrons over X-rays for internal strain measurements is their greater penetrating power. Neutrons can penetrate approximately 1000 times deeper than X-rays in most materials. Cadmium, one of the few elements that strongly attenuate thermal neutrons, is used to make slits that define incident and diffracted neutron beams with rectangular cross sections a few millimeters on each edge. Strain is determined in the small sampling volume where the incident and diffracted beams intersect. Translating the sample with respect to the space-fixed sampling volume yields a profile of how the internal elastic strains and corresponding residual stresses vary with position inside the component.

One method for estimating the residual stresses is to perform computer simulations of the material processing stages by finite-element analysis (FEA). This computational technique lends itself beautifully to experimental verification by neutron diffraction analysis of internal elastic strains.[3,4] The scale of the volume element examined by neutron diffraction (typically a few mm^3) and the precision to which elastic strain can be measured (typically $\pm 1 \times 10^{-4}$) are well matched to the requirements of the simulation method. In this paper we present a comparison of residual stresses near a pin hole in the tang of a stiffener segment as simulated by finite-element analysis at Thiokol Corp. and as measured by neutron diffraction at AECL Research, Chalk River Laboratories.

EXPERIMENTAL

SAMPLE - The specimen was a part of the tang of case segment 1U50715 S/N 032, a lightweight stiffener segment. This segment was flown in the aft position on motor 21A and was used in the forward position on test motor DM-8. A proof test was performed on the case before Flight 21 and before its use in the DM-8 test motor. At the conclusion of the DM-8 test, the segment was subjected to a localized overheating as the result of a failure in the external case cooling system. The case segment was scrapped, following this failure, but specimens were obtained from regions of the case that had not been overheated, and were expected to contain residual stresses typical of a proof-tested component. The principal directions of the original cylindrical case segment are herein denoted as axial (A), hoop (H) and normal (N). The specimen provided to AECL for neutron diffraction analysis had the form of a curved plate with a length in the hoop direction of 49 cm, a length in the axial direction of 23 cm and a thickness near the pin hole of 1.7 cm.

NEUTRON DIFFRACTION - Neutron diffraction measurements were made with the L3 spectrometer at the NRU reactor located at Chalk River Laboratories. A squeezed germanium single crystal monochromator selected an incident beam neutron wavelength of 2.6139 Å, as determined by calibration with a standard reference sample. The incident and diffracted beam cross sections were both set to have height 2 mm and width 1.3 mm. The sample was moved by computer-controlled X, Y and Z translators to a precision of ±0.05 mm to place the sampling volume at the required locations within the specimen. The collimating slits, sample, and translators are shown in figure 1. Strains were measured in the hoop, axial and normal directions as a function of distance from the edge of the pin hole towards the front edge of the tang, and as a function of position through the thickness of the tang, as marked on figure 2. This figure shows a cut through the mid-plane of the pin hole, but in the actual measurements the specimen was left intact. No special surface preparations nor sectioning were required.

Fig. 1 - Photograph of the neutron diffraction apparatus showing the collimating slit holders (A), X, Y and Z translators (B) and the case segment (C) oriented to obtain measurements of the hoop component of strain.

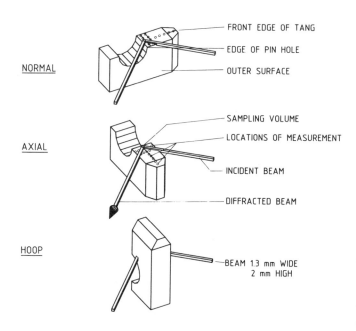

Fig. 2 - Schematic diagram of the tang showing sample orientations for the determination of three components of strain, locations of the neutron measurements and points referenced in the text.

At each position a ^{3}He detector was scanned over a (110) Bragg peak to determine the variation of diffracted neutron intensity with scattering angle, 2θ. The <110> direction in the lattice exhibits an elastic response close to that of bulk material. This property permitted a straightforward comparison between simulated residual stresses and measured elastic strains. By fitting the intensity profile with a gaussian function on a sloping background the angular positions of the diffraction peaks were determined to a typical precision of $\pm 0.012°$. Strains were therefore determined to a typical precision of $\pm 2 \times 10^{-4}$. To obtain a stress-free lattice spacing, d_0, we averaged a number of measurements through the thickness of the specimen 100 mm from the pin hole, where the finite-element calculation indicated that the local residual stresses had decayed to negligible amounts.

FINITE-ELEMENT ANALYSIS - The residual stress profile in the neighbourhood of the pin hole was calculated by a finite-element simulation at Thiokol Corp. The simulation was based on a classical incremental elastic-plastic model. Kinematic hardening was assumed with a bilinear, uniaxial stress/strain curve. The material was assumed to have a 0.2% offset yield strength of 1260 MPa, a Young's modulus of 207 GPa and a plastic modulus of 2800 MPa. The finite-element model was based on worst-case component dimensions with minimal thickness, maximum pin hole diameter and so on. A maximum proof test pressure of 7.5 MPa was assumed. The results of the analysis were provided as contour plots of residual stress in the hoop, axial and normal directions. Values obtained by interpolation from these plots were converted to elastic strains for comparison with the neutron diffraction data by the relation

$$\varepsilon_\alpha = \frac{1}{E}[(1+\nu)\sigma_\alpha - \nu(\sigma_H + \sigma_A + \sigma_N)] \quad (3)$$

where α is one of the principal directions, H, A or N, E is Young's modulus and ν is Poisson's ratio, assumed to be 0.3.

RESULTS

In figure 3 we present the variation of elastic strain through the thickness of the tang at a distance of 3 mm from the edge of the pin hole. The uncertainty in the measurements is shown as a bar of length twice the standard error. The major effect suggested by the simulation is a compressive hoop strain varying from -27×10^{-4} at the outer surface of the tang to -42×10^{-4} at the inner surface. The predicted axial and normal strain components vary little with position through the wall of the tang and are approximately 10×10^{-4} in tension throughout. The neutron diffraction measurements of hoop strain verify that the compression increases towards the inner surface, though it is overestimated in the simulation by about a factor of two. The maximum compressive strain in the hoop direction is $(-23 \pm 2) \times 10^{-4}$. In the axial and normal directions the experimental tensile strain near the outer surface is less than in the simulation, but towards the inner surface the experimental and simulated strains agree. The maximum observed axial and normal tensile strains of $(12 \pm 2) \times 10^{-4}$ occur near the inner surface of the tang.

In figure 4 we present a comparison of the variation of elastic strain through the thickness of the tang at a distance of 7 mm from the edge of the pin hole. The magnitudes of all three strain components are much reduced from those at 3 mm from the edge of the pin hole. Again, the simulation overestimates the actual compressive hoop strain, varying from -7×10^{-4} near the outer surface to -14×10^{-4} at the inner surface. The measurements are consistent with a compression that increases towards the inner surface but are zero, within the experimental precision, near the outer surface. The axial strains are approximately zero throughout the thickness of the tang, both in the simulation and in the experiment. The normal strains from the simulation agree with the neutron measurements, varying from about 3×10^{-4} at the outer surface to about 5×10^{-4} at the inner surface.

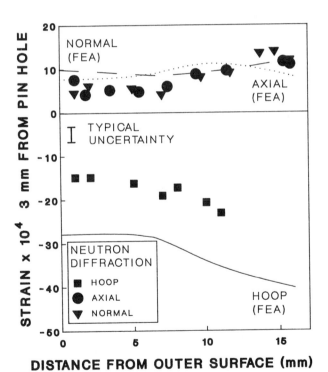

Fig. 3 - Comparison of finite-element simulated elastic strains with neutron diffraction measurements scanning through the wall of the tang, 3 mm from the edge of the pin hole.

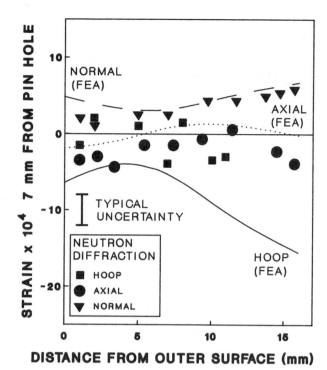

Fig. 4 - Comparison of finite-element simulated elastic strains with neutron diffraction measurements scanning through the wall of the tang, 7 mm from the edge of the pin hole.

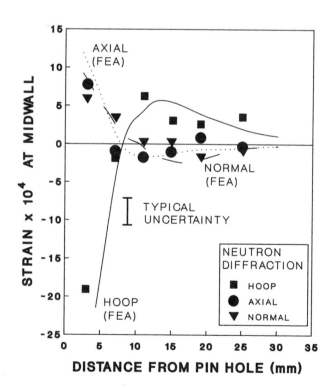

Fig. 5 - Comparison of finite-element simulated elastic strains with neutron diffraction measurements as a function of distance from the edge of the pin hole at midwall.

In figure 5 we show the variation of elastic strain as a function of distance from the edge of the pin hole towards the front edge of the tang. The finite-element simulation indicates a compressive hoop strain exceeding -20×10^{-4} close to the edge of the pin hole, oscillating through a lesser tensile overshoot of 5×10^{-4} at 12 mm from the pin hole and decaying towards zero at greater distances. The axial and normal strains show the inverse pattern with a reduced magnitude. The neutron diffraction results shown for comparison are averages of data obtained within ±1.5 mm of the midwall line at each location between the edge of the pin hole and the front edge of the tang. The only significant discrepancy is close to the edge of the pin hole where, as observed in the previous figures, the simulation overestimates the compression in the hoop direction. However, the agreement in shape of the strain profiles between the finite-element analysis and the neutron diffraction measurements is excellent.

CONCLUSIONS

Neutron diffraction is well-suited for the verification of finite-element calculations of macroscopic strain fields. A simulation of residual stresses, based on worst-case dimensions and properties, was shown to overestimate the elastic strain near a pin hole in the tang of a segment of a solid-fuel booster rocket casing, though the trends of the simulated and measured strain profiles agreed.

ACKNOWLEDGEMENTS

We wish to acknowledge the expert technical assistance of D.C. Tennant, H.F. Nieman, M.M. Potter, G.A. Tapp and L. MacEwen. J.V. Daines of Thiokol Corp. provided valuable discussion and numerical data, contributing greatly to our understanding of the problem.

REFERENCES

1. Holden, T.M., Root, J.H., Fidleris, V., Holt, R.A. and Roy, G., Materials Science Forum, 27/28, 359-370 (1988)

2. I.C. Noyan and J.B. Cohen, "Residual Stress - Measurement by Diffraction and Interpretation" Springer-Verlag, New York (1987)

3. Flower, E.C., MacEwen, S.R. and Holden, T.M., "FEA Predictions of Residual Stress in Stainless Steel Compared to Neutron and X-Ray Diffraction Measurements", 2nd Int. Conf. on Advances in Numerical Methods in Engineering: Theory and Application, Swansea, Wales, (1987) (Lawrence Livermore National Laboratory UCRL-96643).

4. Salinas-Rodriguez, A., Root, J.H., Holden, T.M., MacEwen, S.R. and Ludtka, G.M., Mat. Res. Soc. Symp. Proc. 166, 305-310 (1990)

Residual Stress Characterization in Technological Samples

H.J. Prask
National Institute of Standards and Technology
Gaithersburg, Maryland

C.S. Choi
National Institute of Standards and Technology
Gaithersburg, Maryland
and
ARDEC
Picatinny Arsenal, New Jersey

ABSTRACT

Energy-dispersive neutron diffraction has been developed at the NIST reactor as a probe of sub- and near-surface residual stresses in technological samples. Application of the technique has been made to a variety of metallurgical specimens, and includes the determination of triaxial residual stress distributions as a function of depth in uranium-3/4wt%Ti alloy samples with different thermo-mechanical histories, a number of aluminum-alloy components of military end-items, and induction-hardened steel shafts which are fatigue lifetime test samples for a Society of Automotive Engineers study. Recent results for these shafts will be presented as illustrative of various aspects of the method.

SEVERAL TECHNIQUES ARE AVAILABLE for the nondestructive determination of residual stress, e.g. x-ray diffraction, eddy current, and ultrasonics. Of these, the best established for quantitative characterization is x-ray diffraction which is, generally, a surface probe. Although surface stresses are of prime concern in many engineering applications, certain alloys present serious difficulties for the x-ray technique because of large grain size, surface contamination, or texture.

Neutron diffraction closely parallels x-ray diffraction in methodology and analytical formalism. However, because neutrons interact with nuclei and x-rays with electrons, neutrons are typically about a thousand times more penetrating than x-rays in the wavelength range for diffraction ($0.07 \leq \lambda \leq 0.4$ nm). In addition, different elements exhibit significantly different relative scattering powers for neutrons and x-rays. A utilization of the unique aspects of neutron diffraction for the "depth-profiling" of texture was demonstrated by us in 1979 [1]. Since the initial measurements in 1982 [2] a number of tests and applications of the neutron diffraction technique to sub-surface residual stress determination have been reported [3,4].

In the present paper we review fundamentals of the technique and describe the application of neutron diffraction to the determination of subsurface triaxial stress distributions in two induction-hardened steel shafts. The results are compared with analogous x-ray measurements in which layer removal was employed.

METHODOLOGY

GENERAL CONSIDERATIONS - In both x-ray and neutron diffraction determination of residual stress, what is measured is strain which is manifested by changes in the distance, d, between atomic planes in the sample. A unique advantage of neutron diffraction arises from the different relative scattering cross-sections and penetration relative to x-rays. This is illustrated in Table 1 in which $t_{1/2}$, the thickness at which half the beam intensity, in transmission, is lost through scattering and absorption processes, is listed for selected metals. The values are based on cross-sections from standard references and the difference in wavelengths used for neutrons (0.108 nm) and x-rays (0.154 nm) is not significant.

Table 1. Neutron and X-ray Comparison

Element(At.No.)	$t_{1/2}$(X-rays)	$t_{1/2}$(Neutrons)
Al (13)	0.0530 mm	71.0 mm
Ti (22)	0.0076	15.9
Fe (26)	0.0027	6.1
Cd (48)	0.0035	0.057
W (74)	0.0021	6.5
U^{238} (92)	0.0015	13.6

It should be mentioned that the $t_{1/2}$ values in Table 1 do not represent the depth of penetration in a residual stress measurement. This is dependent on a number of factors such as source intensity, coherent scattering cross-section, and beam spot size. However, the $t_{1/2}$ values clearly show that neutrons in the normal diffraction wavelength range are several orders-of-magnitude more penetrating than x-rays. Also, the penetration does not decrease monotonically with atomic number as with x-rays, but is a random function of atomic number. It is clear from Table 1 that, for example, depleted uranium is a good material for neutron examination [3]. In contrast, no reliable determination of residual stresses has been obtained with conventional x-ray techniques for this material.

The properties of neutrons suggested the possibility of measuring subsurface residual stresses in metallurgical samples employing the special collimation and the scattering geometry shown schematically in Figure 1. A scattering angle, Ω, of ~90° is used to minimize the examined differential volume, ΔV. The perspective view in Figure 1 shows ΔV defined by two rectangular apertures in an absorbing material, e.g. cadmium; however, the apertures could be any shape including circular. The plan view in Figure 1 indicates how the sample can be translated in the beam so that ΔV can be examined as a function of depth.

The strains from which residual stresses are inferred are obtained from measured d-spacings through Bragg's Law:

$$\lambda = 2d(hkl)\sin \Omega/2$$

where λ is wavelength, $d(hkl)$ is the separation of atomic planes with Miller indices hkl, and Ω is stepped - usually with the sample position stepped by $\Omega/2$ - and sharp resonances in scattered intensity are observed at scattering angles where the Bragg condition is fulfilled. Precise determination of $\Omega(hkl)$, the peak position, yields $d(hkl)$ directly. Although success has been achieved with this mode of measurement, as reviewed in reference 4, some problems have been encountered with highly attenuating or highly textured samples. Here, texture means the existence of preferred crystallographic orientation of grains or crystallites relative to a coordinate system fixed in the sample.

With reference to the plan view in Figure 1, it is clear that as Ω is varied, path length to and from ΔV changes and intensity as a function of Ω is possibly distorted leading to a false shift in $d(hkl)$. Similarly, gradients in preferred grain orientation in the sample over the changing beam-in/beam-out paths can produce intensity variations which shift the apparent $\Omega(hkl)$. Since the strains, $\Delta d/d_0$ are on the order of 0.0001 (where d_0 is the d-spacing in the absence of residual stress), a small anomalous shift nullifies the stress measurement.

ENERGY-DISPERSIVE NEUTRON DIFFRACTION - In our measurements, we have made use of the fact that Bragg-condition resonances can also be observed at fixed Ω with varying wavelength. With the scattering angle fixed, changing attenuation and texture gradients can be less distortive. In addition, the examined volume ΔV, remains exactly the same throughout each scan.

The instrument used for energy-dispersive neutron diffraction (EDND) is called a triple-axis spectrometer. Crystals of known d-spacing are placed before and after the sample; the Bragg relation is then used to select and step the wavelength incident on the sample. In principal, the analyzer crystal - which we step at the identical wavelength as the monochromator - is not needed. However, utilization of the analyzer significantly enhances instrumental resolution. Both EDND and angle-dispersive neutron diffraction benefit from peak profiles which are Gaussian in shape and straightforward to analyze with least-squares curve-fitting techniques.

In our system, pyrolytic graphite crystals were used for the monochromator (002 plane) and analyzer (004). The collimation employed was 50'-x-x-40' from source to analyzer, with a resultant resolution $\Delta\lambda/\lambda = 0.0073$ at $\lambda = 0.2692$ nm, where the "x" denotes the Cd-defined apertures. The Cd absorbers were cut with 1x5, 2x2, 2x5 or 6x6 mm² apertures for reasons discussed below.

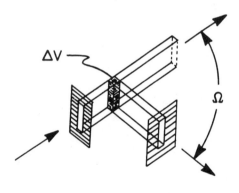
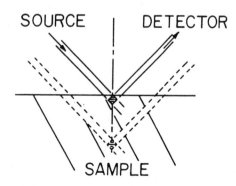

Figure 1. Schematic of how absorbing masks are used with Bragg's Law to define the examined volume, ΔV (upper); and plan view of ΔV translation through the sample (lower) in the neutron diffraction measurements.

STRESS-STRAIN RELATIONS - The relation between stress and strain applicable to diffraction measurements has been presented, for example, by Evenschor and Hauk [5].

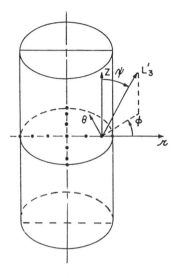

Figure 2. Coordinate system and possible measurement mesh for the neutron diffraction measurements. The solid circles represent the centers of the ΔVs examined.

With reference to Figure 2, r, θ, and z are specimen-fixed axes, and the strain $\epsilon'_{\phi\psi}$ is measured along $\vec{L}'_3$; then

$$\begin{aligned}\epsilon'_{\phi\psi} &= (d_{\phi\psi}-d_0)/d_0 \\ &= (\epsilon_{11}\cos^2\phi + \epsilon_{12}\sin2\phi + \epsilon_{22}\sin^2\phi)\sin^2\psi \\ &\quad + (\epsilon_{13}\cos\phi + \epsilon_{23}\sin\phi)\sin2\psi + \epsilon_{33}\cos^2\psi\end{aligned} \quad (1)$$

where $d_{\phi\psi}$ is the lattice spacing along $\vec{L}'_3$ and d_0 is the unstressed lattice spacing. The stresses are related to the measured strains through

$$\begin{aligned}\epsilon'_{\phi\psi} &= 1/2 S_2(hkl)[\sigma_{11}\cos^2\phi\sin^2\psi + \sigma_{22}\sin^2\phi\sin^2\psi \\ &\quad + \sigma_{33}\cos^2\psi + \sigma_{12}\sin2\phi\sin^2\psi + \sigma_{13}\cos\phi\sin2\psi \\ &\quad + \sigma_{23}\sin\phi\sin2\psi] + S_1(hkl)[\sigma_{11}+\sigma_{22}+\sigma_{33}].\end{aligned} \quad (2)$$

The $S_i(hkl)$ are diffraction elastic constants ("XEC") for the (hkl) reflection which, in general, depend on the material and the reflection examined. For an elastically isotropic solid the XEC are given by

$$1/2 S_2(hkl) = (1+\mu)/E \text{ and } S_1(hkl) = -\mu/E \quad (3)$$

where μ, E are Poisson's ratio and Young's modulus, respectively.

Since the determination of residual stress in engineering samples by means of Eq. 2 depends directly on measurement of strain values, a precise value for the unstressed d-spacing, d_0, is essential. We have been successful in determining unstressed lattice parameter from neutron measurements on powders in some cases; however, powder samples identical to the specimen under study often cannot be obtained. Alternatively, we utilize the overall equilibrium conditions required by elasticity theory to determine d_0. That is, since the body is static with no external force applied, residual stresses normal to any plane must balance such that in cylindrical geometry:

$$\int \sigma_{zz} r\,dr\,d\theta = 0, \quad \int \sigma_{\theta\theta} dr\,dz = 0 \quad (4)$$

and at any surface the stress orthogonal to that surface must vanish. Stresses inferred from measured strains can be adjusted, by varying d_0, to fulfill the equilibrium conditions, which - in turn - yields a d_0 value.

MEASUREMENTS AND RESULTS

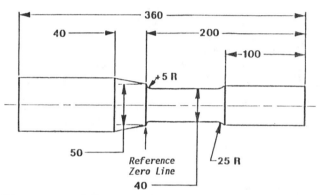

Figure 3. SAE biaxial fatigue test "axle". The 100 mm length on the right is the stationary portion in fatigue test cycling. The diameters at the left and right ends of the axle are 63.5 and 45 mm, respectively.

The Society of Automotive Engineers Fatigue Design and Evaluation Committee has been conducting a long-term program aimed at the development of a predictive capability for fatigue life. The first phase of this program examined unhardened SAE 1045 steel shafts subjected to a variety of torsion and/or bending strains [6]. In Phase II of this program the multiaxial fatigue behavior of induction-hardened and tempered SAE 1045 notched shafts is being examined. In this phase a more comprehensive, integrated engineering approach is being applied to the problem than in Phase I. It includes materials properties characterization, finite element analysis prediction of stress/strain, residual stress characterization, fatigue-life measurement, and fatigue lifetime prediction. Some depth-profiling of residual stresses has been done by x-ray diffraction with layer removal [7]; however, the nondestructive character of neutron diffraction offers an excellent means for both characterizing the residual stress distributions and in testing the x-ray layer-removal approach.

The geometry of the component-like test "axles" is shown in Figure 3. Two samples were studied with neutron diffraction: an axle not subjected to any strains, and an axle which had been subjected to bending and torsion fatigue cycling to one-half the fatigue life (based on fatigue cycling to failure of essentially identical specimens). Induction hardening of the surface region in these samples introduces complications to what would have been - except for the relatively high attenuation of thermal neutrons by iron - a straightforward stress determination. In Figure 4, measured hardness versus depth is shown for several positions on the unfatigued shaft [8]. The near-surface region (0 - 3 mm depth) is predominantly tempered martensite, changing to a mixture of pearlite/ferrite at ~10 mm depth. To determine the reference d_0's for the different regions, a third axle was cut by EDM into a 3 mm thick "martensite" ring which included the surface, and a shaft-centered 20 mm diameter pearlite/ferrite plug, each 1 cm wide. It should be mentioned that in the utilization of Eqn. 4 to determine d_0, absolute values of d_0 are not required, only an internally consistent set for the conditions of the experiment.

requiring balance of axial stresses in the central r-θ plane of each piece. XEC determined by Prevey [9] for the (211) reflection of SAE 1050 steel were used. Current theory indicates that the XEC for the (211) and (110) reflections are identical [10].

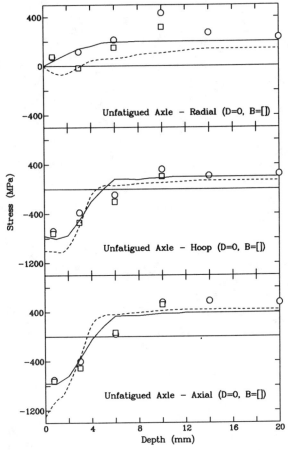

Figure 5. Neutron diffraction determined residual stresses for the unfatigued axle compared to the finite element analysis [11]. The dashed line is the FEA result at 5 mm from the *reference zero line*; the neutron "B" values are at 10 mm[1].

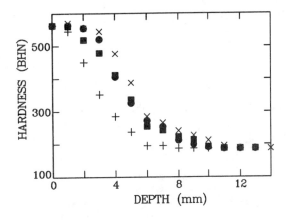

Figure 4. Hardness vs. depth for the induction hardened shafts [8] for various distances for the reference zero line: ● = 10 mm ("B"), X = 25 mm, ■ = 50 mm (interpolated, "D"), + = 110 mm.

For all of the measurements the (110) reflection of steel was employed at a scattering angle of 80° from the sample. The d_0's for the martensite and pearlite/ferrite structures were obtained from the ring and plug pieces, respectively, with 2x2 mm² apertures and the macroscopic equilibrium conditions for axial stresses. Strain measurements were made at a total of seventeen points along two mutually perpendicular diameters - eight points in the ring - from which d_0(ring)= 0.20323 nm and d_0(plug)=0.20303 nm were obtained by

Because of the high neutron attenuation of steel, 6x6 mm² apertures were used to determine d-spacings at r=0 and r=6 mm in the unfatigued axle at position "D" (50 mm from the *reference zero line*[1] indicated in Figure 3). Other measurements on the two 20 mm radius shafts were made with 2x5 mm² apertures oriented either horizontally or vertically for measurements at r= 10, 14, and 17 mm, and with 1x5 mm² apertures at r=19.2 mm; the 5 mm dimension was parallel to the cylindrical axis

---

[1]It should be noted that the "B" and "D" positions used here are close to but not exactly those used in the originally reported hardness measurements. In this paper, we define "B" and "D" with respect to the *reference zero line* of Figure 3.

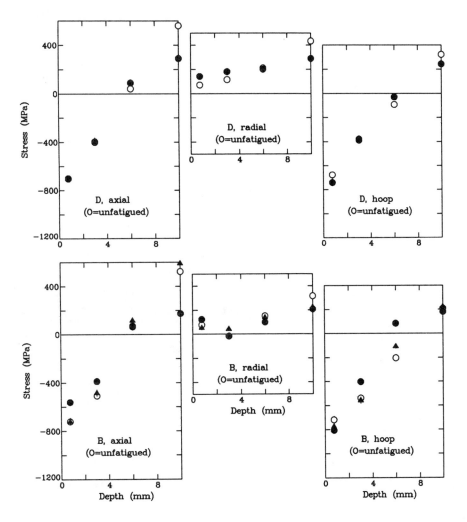

Figure 6. Neutron diffraction determined residual stresses at the 10 mm ("B") and 50 mm ("D") positions for the fatigued (●) and unfatigued (○) samples. The triangles correspond to stresses at a radius 90° to the bending axis.

of the shaft in either orientation. Data were obtained along single radii at positions "B" (= 10 mm from the *reference zero line*) and "D" in the unfatigued shaft; at "B" and "D" on the radii parallel to the bending direction, and at the "B" position at 90° from the bending direction in the fatigued axle. All d-spacing data were normalized to Fe(110) of an iron powder which was scanned for each change of apertures.

A test of the ring-plug d_0's was made using the unfatigued axle data at the "D" position. Using the hardness vs. depth results of Langner [8], d_0's for each depth examined by neutrons can be inferred, with the ring-plug d_0's at the extremes. The hardness at the "D" (= 50 mm) position was obtained by a linear interpolation of the 25 and 95 mm measured hardness values. If the approach is correct, axial stresses measured in the r-θ plane at "D" should balance with no parameter adjustment for appropriately chosen area segments for each point of measurement (e.g. the 2x2 mm² beamspot cross-section centered at a depth of 3 mm is assumed to represent the full 2 mm wide ring of area from r=16 to 18 mm of which it is a small part, etc). The axial stresses balance to 10 MPa, which is to say that perfect balance would be achieved if the axial stress determined at each of the six measurement points was shifted by 10 MPA. Since the standard deviation on each measured axial stress value is typically ~20 MPa, this balance-of-stresses represents an excellent confirmation of the approach.

Ochsner [11] has made a finite element analysis (FEA) of an induction hardened steel shaft to obtain the residual stress distribution. The results are shown in Figure 5 along with the neutron-diffraction-determined stresses at positions "B" and "D" in the unfatigued shaft. The effect of fatigue on the residual stress distributions is illustrated in Figure 6 where results for the two shafts are shown at corresponding positions.

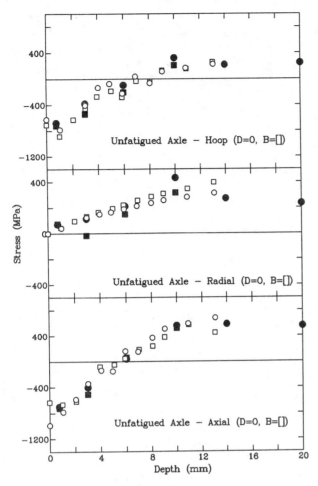

Figure 7. Comparison of the nondestructive, neutron diffraction determined residual stresses (solid points) and x-ray diffraction determined stresses using layer removal [7] for the unfatigued axle.

In Figures 7 and 8 a comparison is made of the results of the nondestructive neutron determination and the x-ray diffraction measurement of residual stress employing layer removal [7]. It should be pointed out that x-ray and neutron measurements were at the same positions on the unfatigued axle, whereas two x-ray measurements on the fatigued axle bracketed the neutron "B" position.

DISCUSSION

Overall, the agreement between the nondestructive neutron results and the x-ray diffraction results, obtained employing layer removal, is excellent. Although layer removal for symmetric specimens has been known to be a reliable technique for subsurface stress determination with x-rays, the agreement obtained in this case to a depth of 13 mm seems remarkable. The only point at which there is a significant difference occurs at a depth of 3 mm for the fatigued axle at the "B" position for all stress components (Figure 8), and at the 3 mm depth for the radial stress component at the "B" position in the unfatigued axle. The fact that the neutron results for σ_{rr} at 3 mm depth for both the fatigued and unfatigued axles are slightly compressive at the "B" position, and that these results are in qualitative agreement with the finite element calculation, supports the correctness of the neutron determination at this point.

The FEA results are in reasonably good agreement with the stress measurements for the unfatigued axle. However, except for the radial stress at 3 mm depth, the measured stresses seem to show much less "B"- vs. "D"-position difference than the FEA predicts. No FEA results are yet available for the fatigued-axle

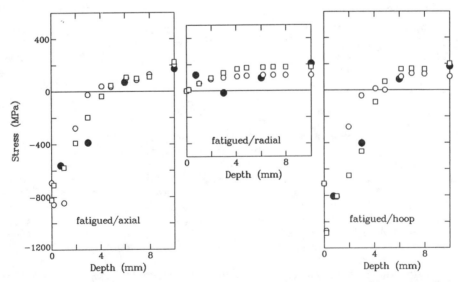

Figure 8. Comparison of the nondestructive, neutron diffraction determined residual stresses (solid points) and x-ray diffraction determined stresses using layer removal [7] for the fatigued axle. The x-ray stress values were obtained at 5 mm on either side of the center of the neutron measurements.

case. Experimentally, the measurements at the "D" position show very little stress relaxation with fatigue except at 10 mm depth for σ_{zz}. In contrast, except at this depth, there seems to be a tendency for the axial and hoop stresses to become more tensile with fatigue at the "B" position. It is worth noting that at the "B" position, the stresses measured with neutrons at 90° to the bending direction are very close in magnitude to the unfatigued stresses, as expected.

ACKNOWLEDGEMENTS

The authors are indebted to Mr. T. Cordes and Mr. M. Langner of John Deere Co. and Mr. P. Prevey of Lambda Research Co. for many helpful discussions relating to this work.

REFERENCES

1. C.S. Choi, H.J. Prask, and S.F. Trevino, J. Appl. Cryst. 12, 327-331 (1979).
2. A. Allen, C. Andreani, M.T. Hutchings and C.G. Windsor, NDT International, 249-254 (Oct. 1981); Krawitz, A.D., Brune, J.E. and Schmank, M.J. in "Residual Stress and Stress Relaxation", edited by E. Kula and V. Weiss (Plenum Press, New York and London, 1982) pp. 139-155; Pintschovius, L., Jung, V., Maucherauch, L., Schäfer, R. and Vöhringer, O., ibid., pp. 467-482.
3. H.J. Prask and C.S. Choi, in "Residual Stress in Design, Process and Materials Selection", edited by W.B. Young (ASM International, 1987), pp. 21-26, and references cited.
4. See for example, "Neutron Scattering for Materials Science", edited by S.M. Shapiro, S.C. Moss, and J.D. Jorgenson (MRS, Pittsburgh, 1990), pp. 281-344, and references cited.
5. P.D. Evenshor and V. Hauk, Z. Metallkde. 66, 167-8 (1975).
6. "Multiaxial Fatigue: Analysis and Experiment", (SAE AE-14), edited by G. Leese and D. Socie, (SAE Inc., 1989).
7. P. Prevey, SAE Fatigue Design & Evaluation Committee Minutes, April 1989.
8. M. Langner, SAE Fatigue Design & Evaluation Committee Minutes, October 1988.
9. P. Prevey, Lambda Research, Inc., Cincinnati, OH, private communication.
10. F. Bollenrath, V. Hauk, and E. Muller, Z. Metallkde. 58, 76-82 (1967).
11. J.K. Ochsner, John Deere Product Engineering Center, Waterloo, IA, unpublished.

In-Process Control and Reduction of Residual Stresses and Distortion in Weldments

K. Masubuchi
Massachusetts Institute of Technology
Cambridge, Massachusetts

ABSTRACT

The emphasis of the efforts by the Welding Research Group at the Department of Ocean Engineering, M.I.T. during the last several years has been for developing technologies of reducing these stresses and distortion through in-process control. The studies performed thus far include (1) reduction of joint mismatch during butt welding in steel and aluminum, (2) reduction of residual stresses and distortion in weldments in high-strength steels, (3) reduction of radial distortion and residual stresses in girth-welded pipes, and (4) reduction of longitudinal bending distortion of built-up beams. This paper presents results obtained in these studies with an emphasis on the first two studies.

A MAJOR PROBLEM associated with arc welding is that related to residual stresses and distortion. Because a weldment is locally heated by the arc, complex thermal stresses occur during welding, and residual stresses and distortion remain after welding is completed. These stresses and distortion cause complex consequences, most of which are detrimental, during fabrication and service. They include the formation of joint mismatch during welding, cracking, and premature failures of welded structures [1]. The Welding Research Group at the Department of Ocean Engineering of the Massachusetts Institute Technology has performed research for many years on various subjects related to residual stresses and distortion in welded structures. The emphasis of the M.I.T. research during the last several years has been on developing technologies of reducing these stresses and distortion through in-process control. The studies performed thus far include the following:

(1) Reduction of joint mismatch during butt welding in steel and aluminum

(2) Reduction of residual stresses and distortion in weldments in high-strength steels

(3) Reduction of radial distortion and residual stresses in girth-welded pipes

(4) Reduction of longitudinal bending distortion of built-up beams.

This paper summarizes results obtained in these studies. Because of the page limitation of this paper, the emphasis of this paper is placed on the first two studies.

BASIC CONCEPT OF IN-PROCESS CONTROL OF RESIDUAL STRESSES AND DISTORTION

The reason why real-time control is important for reducing residual stresses and distortion can be understood by studying mechanisms of their formation. Figure 1 shows schematically how a rectangular plate deforms when arc welding is performed along its upper longitudinal edge. Since temperatures are higher in regions near the upper edge, these regions expand more than regions near the lower edge causing the upward movement of the center of the plate, δ, as shown by Curve OA. The most important stress component is the longitudinal stress, σ_x. Stresses in regions near the weld are compressive, because thermal expansions in these regions are restrained by the surrounding metals at lower temperatures. Since the temperatures of the regions near the weld are quite high and yield stresses of the material are low, compressive plastic strains are produced in these regions. When welding is completed and the plate starts to cool, it deforms in the opposite direction. If the material was completely elastic during the entire period of the heating and cooling cycle, the plate would deform as shown by Curve OAB'C' returning to its initial shape with no residual distortion. However, this does not happen during welding a real material, be it steel, aluminum, or titanium. As a result of the compressive plastic strains produced in regions near the upper edge, the plate continues to deform after passing its initial shape, as shown by Curve OABC, resulting in the negative final distortion, δ_f, when the plate cools down to its initial temperature.

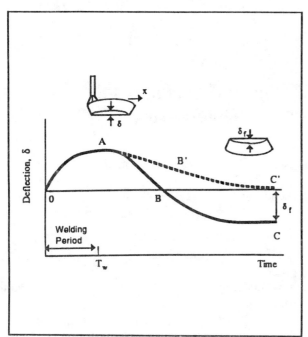

Figure 1 Transient Deformation of a Rectangular Plate During Welding

The most effective way of reducing distortion is to control the formation of plastic strains produced in regions near the weld. The difficulty here is that the necessary control must be made during welding. If the control performed is correct, the final distortion will be reduced. If the control is incorrect, on the other hand, the final distortion will be increased. In order to perform correct controls consistently, one must have the following capabilities:

(1) <u>Prediction capability</u>. One must have a proper capability of predicting, by analysis, prior experiments, and/or experience, (a) how the weldment being studied deforms and (b) how to perform proper controls to change the distortion being considered.
(2) <u>Sensing capability</u>. One must also have a proper device or devices for sensing if what should happen is actually happening.
(3) <u>Control capability</u>. If one finds that what is actually happening is different from what is supposed to happen, it is important that he/she has capability of making necessary changes, in real-time if needed.

Efforts have been made at M.I.T. for improving these capabilities. Regarding the prediction capability, a series of computer programs have been developed including (a) simple one-dimensional programs which analyze only the stress component parallel to the weld line or the longitudinal stress and (b) finite-element programs capable of analyzing more complex stress fields.[1] Regarding the sensing capability, efforts have been made for improving techniques for measuring out-of-plane distortion. They include: (a) a laser interferometer, (b) a laser vision system, and (c) a mechanical system capable of measuring radii of curvature in four directions around a measuring point [2].

REDUCTION OF FORCES ACTING ON TACK WELDS IN A BUTT JOINT

A research program was performed for the Department of Energy with an objective of minimizing forces acting on tack welds in a butt joint. The uneven temperature distribution caused by the arc produces complex transient thermal stresses resulting in mismatch of parts to be joined unless they are securely held together. The joint mismatch that occurs during butt welding, which is shown in Figure 2-(a), can be explained by combining the information given in Figure 1. Suppose that a butt weld is being made with no tack weld, as shown in Figure 2-(a), each of the two parts being joined behaves as shown by Curve OABC in Figure 1. If deformations in regions near the welding arc when they start to cool are near Point A (or somewhere between Points O and B), the finishing end of the joint will open. This phenomenon normally occurs during gas metal arc and submerged arc welding of steel plates. On the other hand, when a joint is welded with covered electrodes, deformations of regions near the weld when they start to cool may be somewhere after Point B, resulting in closing of the finishing end. This type of distortion is often called "rotational distortion" [1].

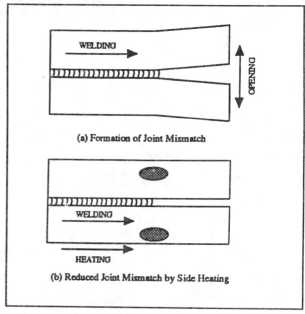

Figure 2 Mechanism of Formation of Joint Mismatch in a Butt Weld and Reduction of Mistmatch by Side Heating

A common method of coping with the rotational distortion is to hold the joint with tack welds. This can be done relatively easily in manual welding of small parts. In the case of automatic welding, however, dealing with the rotational distortion becomes a complex problem. When welding is performed by a robot, for example, tack welds must be made by a human welder, thereby requiring additional manpower and cost. On many occasions, tack welds are performed by an inexperienced person, resulting in less than perfect welds. Also, tack welds, even though they are perfectly made, act as major hazards during the subsequent root pass welding. It is difficult to completely melt tack welds during the root pass welding, thereby causing lack of penetration and other types of defects [3]. In submerged arc welding a long butt joint of thick plates, forces acting on tack welds are so great that they often break during welding. In fact, Japanese shipbuilders experienced longitudinal cracking of the finishing end when they first introduced one-side submerged arc welding of large ship plates [4].

Chang, Park, and Miyachi performed experimental and analytical studies for reducing forces acting on tack welds during butt welding [5-7]. The basic idea used by Chang was to reduce the joint mismatch or the rotational distortion by side heating, as schematically shown in Figure 2-(b). By performing side heating while welding is being performed it may be possible to produce additional thermal stresses that can counteract those produced by welding. It is important, however, that the additional heating not produce additional residual stresses.

Figure 3 shows the experimental set up used by Chang [5]. Instead of using tack welds for holding plates to be joined, a semi-circular ring was attached to each end of the joint. The rings were instrumented with strain gages in order to measure changes of deformation at these ends. Although the rings were welded to the plates in experiments, these rings can be designed in such a way that they can be clamped to plates to be welded in actual fabrication. Efforts were made to reduce the opening of the finishing end by altering the thermal pattern in the weldment during welding. Two oxygen torches were mounted on a frame with a welding head so that the side heating system could be moved along with the welding arc. The position of the heating system relative to the welding head could be adjusted in three directions, x, y, and z, as shown in Figure 3, in order to control the side heating procedure. The system was used mainly for controlling the joint mismatch or the rotational distortion in a steel weldment.

<u>Results on Steel Weldments</u>. Figure 4-(a) shows typical results obtained on a low-carbon steel weldment, 36 inches (914 mm) long, 24 inches (610 mm) wide, and 0.5 inch (12.7 mm) thick, with no side heating. Very soon after welding commenced, the starting end began to shrink. Note that when closing of a joint occurs, strains measured on gages attached on the outer surface of the ring would be tensile. On the other hand, the finishing end first opened as welding progressed, and it began to shrink after welding was completed. Almost all of the difference between the movement of the starting end and that of the finishing end was produced during welding. Welding was completed in 165 seconds, but it took approximately 1,800 seconds (30 minutes) before forces acting on the rings were fully developed.

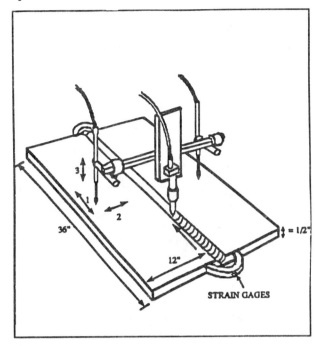

Figure 3 Side Heating System Used

Figure 4-(b) shows results obtained on a steel weldment with side heating. It is clear that the amount of joint opening decreased significantly by use of the side heating. The dotted line in the figure is an ideal case in which no joint mismatch occurs. Results shown in Figures 4-(a) and -(b) illustrate that the deformation at the starting end is little affected by the side heating. This indicates that the measurement at the starting end can be used as a control. In other words, the joint mismatch can be minimized as long as the strain measurements on the finishing end are similar to those obtained on the starting end.

An analytical study was made to determine optimum conditions for side heating using a one-dimensional program that analyzes the longitudinal stress only. It has been found that the most desirable side heating is to heat wide regions away from the weld to moderate temperatures (around 200°F or 93°C) to accomplish the following:

(a) The side heating should produce thermal stresses large enough to counteract those produced by welding.

(b) The side heating should not produce additional residual stresses.

In order to develop a strategy for most effectively controlling the joint mismatch, a series of experiments were performed to study effects of torch movements in the x-, y-, and z-directions. It has been found that forces acting on the ring at the finishing end, or a simulated tack weld at the finishing end, can be significantly reduced by a proper side heating. For example, the maximum opening force observed on a ring attached to the finishing end was reduced from 1,125 pounds without side heating (welding only) to only 105 pounds with side heating.

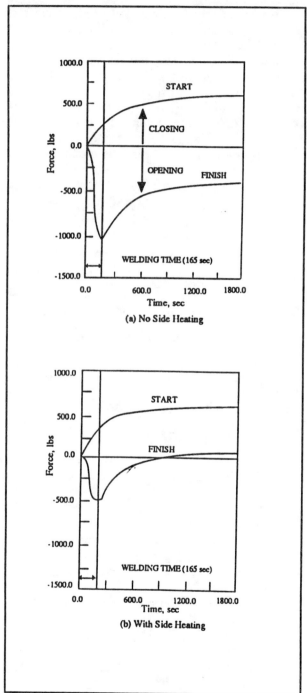

Figure 4 Forces Acting on the Rings Attached to Starting and Finishing Ends of Butt Joints in Low-Carbon Steel

Results on Aluminum Welds. A limited amount of work has been performed on aluminum welds by Park [6]. First, experiments were performed to study effects of side heating on forces acting on rings attached to both ends of aluminum butt welds 36" x 24" x 0.5" in size. Results were very disappointing. For example, the maximum opening force observed on a ring attached to the finishing end increased from 300 pounds on a weld without side heating to 700 pounds on a weld with side heating. This is probably due to combined effects of the following:

(a) Compared with steel, aluminum has a large heat conductivity (approximately 5 times of that of steel). Therefore, the heat spreads much more rapidly in aluminum than in steel. In order to produce by side heating thermal stresses large enough to counteract those produced by welding, we must have uneven temperatures caused by both the welding arc and the side heating. In an aluminum weld, the heat spreads so rapidly that temperature distributions caused by the welding and the side heating cannot be well separated.

(b) Compared with steel, aluminum has a large coefficient of linear thermal expansion (approximately 3.5 times of that of steel). Therefore, the best method for reducing distortion in an aluminum weld is to lower temperatures, not to increase them by additional heating.

It was decided to study effects of forced cooling on the joint mismatch during welding an aluminum butt joint. Since it was difficult to have a heat sink that can travel with the welding arc, it was decided to cool the joint before welding to keep the cooling system operating during the entire welding period. Crushed dry ice particles were used as the coolant. The cooling was done in regions near the weld. Then the maximum force observed on a ring attached to the finishing end decreased from 300 pounds (135 kg) without cooling to merely 30 pounds (13.5 kg) with cooling. The results show that the key for reducing the joint mismatch (and perhaps residual stresses also) in an aluminum weld is to keep the temperatures in the weldment as low as possible.

Development of Real-Time Control System. Miyachi has developed a real-time control system, based on experimental results obtained by Chang, for minimizing forces acting on a tack weld [7].

REDUCTION OF RESIDUAL STRESSES AND DISTORTION IN HIGH-STRENGTH STEEL WELDMENTS

Various types of high-strength steels are increasingly used for producing welded structures with reduced weight and improved performance. A welding problem here is that related to residual stresses and distortion. Since yield stresses of these steels are high, there is always a possibility of producing very high residual stresses in some locations near the weld, including the end of a repair weld and regions near structural discontinuities (such as the end of a fillet weld connecting a flat plate and a stiffener). High transient thermal stresses during welding and residual stresses may cause cracking and premature fractures during service. Efforts have been made to study means for reducing residual stresses and distortion in weldments in HY-100 and HY-130 steels, quenched and tempered steels with minimum yield strengths of 100 ksi (689 MPa) and 130 ksi (896 MPa), respectively. HY-80 and HY-100 steels have been widely used for submarine hulls, while HY-130 steel is being considered as the major material for hull structures of future submarines. The study was supported by the M.I.T. Sea Grant College Program that received funds from the National Sea Grant College Program, the National Oceanic and Atmospheric Administration.

Presented here is a brief summary of the experimental and analytical investigations performed by Bass and Vitooraporn [8, 9]. Regarding the weldment type, a bead-on-edge weld as shown in Figure 1 was used. The specimen size was 18" (460 mm) long, 5.5" (140 mm) wide, and 0.5" (12.7 mm) thick. Welding was done with a gas metal arc process. A side heating system similar to that used by Chang (see Figure 3) was used. The basic idea was to see whether residual stresses and distortion could be reduced by side heating wide regions somewhere away from the weld to moderate temperatures during welding. A series of experiments were performed on welds in three types of materials: low-carbon steel, HY-100, and HY-130 steels.

Experimental data were obtained on (a) temperature changes during welding, (b) transient thermal stresses during welding, as well as residual stresses after welding is completed, and (c) changes of deformation during welding as well as distortion after welding is completed. Analytical studies using the finite element method also were made, and they had good agreement with the experimental data. It has been found that significant reduction of residual stresses and distortion ranging from 17 to 39% were achieved in all steels studied. Figures 5-(a) and -(b) show a few examples of experimental results. Shown here are relationships between the transverse distance from the weld and longitudinal residual stresses in weldments in low-carbon steel and HY-100 steel. Results obtained on HY-130 steel weldments were similar to those obtained on HY-100 steel weldments shown in Figure 5-(b). A finite element analysis of transient thermal stresses produced in regions near the welding arc has revealed that regions up to approximately 0.45 inch (11 mm) from the weld line experience plastic deformation during welding, but the size of the plastic zone decreases to approximately 0.3 inch (7.7 mm) when side heating is applied during welding.

REDUCTION OF RADIAL DISTORTION AND RESIDUAL STRESSES IN GIRTH-WELDED PIPES

Studies have been made to develop techniques for reducing radial distortion and residual stresses produced by girth welding of pipes. The studies were performed for the Department of Energy.

In the first study performed by DeBiccari, a turnbuckle was used to provide an additional restraint to a specimen (see Figure 6) [10]. Forces generated by the turnbuckle were transmitted to the specimen through two semi-circular shoes so that various locations along the girth were subjected to varying degrees of restraint. The turnbuckle was instrumented with strain gages in order to monitor changes of restraining forces during welding. The inner diameter of the pipe was 12 inches (305 mm), and the wall thickness was 5/16 inch (8 mm). The results obtained may be summarized as follows:

(1) <u>Distortion Shape</u>. The amount of radial contraction is the largest near the weld, and it greatly decreases as the

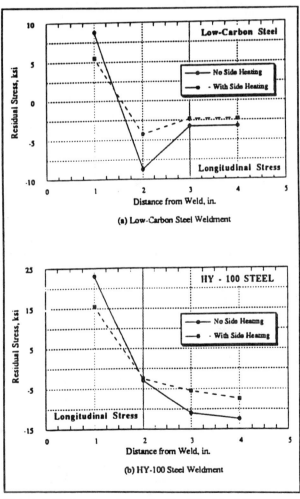

Figure 5 Reduction of Longitudinal Residual Stresses by Side Heating

longitudinal distance from the weld increases.

(2) **Effectiveness of Additional Restraint**. The additional restraint provided by the turnbuckle system decreases the radial contraction very effectively. As the angle q increases, the degree of restraint by the shoe decreases; therefore, the radial contraction increases. DeBiccari also found that restraining forces measured by strain gages mounted on the turnbuckle decreased momentarily when the welding arc passed areas where the turnbuckle touched the pipe. These reductions of restraining forces are believed to be caused by the expansion of the pipe due to the welding heat. He also found that residual stresses measured on the restrained pipe were generally lower than those obtained on the pipe with no additional restraint.

In the second study performed by Barnes, a restraining system using six hydraulic pistons was constructed and used in order to provide constant amount of restraint to a specimen even while it expands due to the welding heat [11]. The system was designed in such a way that it can be easily assembled in a pipe and it also can be easily disassembled after welding is completed.

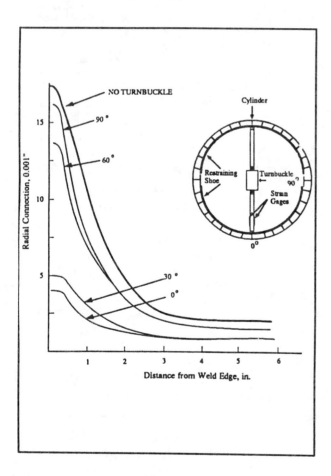

Figure 6 Reduction of Residual Distortion of a Grith-Welded Steel Pipe by Application of Internal Pressure Using an Instrumented Turnbuckle System

REDUCTION OF LONGITUDINAL DISTORTION OF BUILT-UP BEAMS

A very widely used welding fabrication is to manufacture T-beams and I-beams by fillet welding a web to a flange or flanges. Since the weld line is normally away from the neutral axis of the beam, the longitudinal shrinkage in regions near the weld causes longitudinal bending distortion that can be significant in fabricating a long beam. One can imagine from Figure 1 that if the initial temperatures of the web and the flange differ in the right amount, the final distortion of the built-up beam could be minimized. This method can be called the "differential heating." Experimental and analytical studies were performed for reducing longitudinal bending distortion of built-up beams by this technique. Since the work is older than those presented above, and the results are already included in Reference [1], only a brief a summary is presented here.

Serotta conducted a series of experiments to investigate how differential heating reduces the longitudinal distortion produced during welding fabrication of a T-beam in 5052-H33 aluminum alloy [12]. A web plate, 48" x 6" x 0.5" (1,220 x 152 x 12.7 mm), was fillet welded to a flange plate, 48" x 4" x 0.5" (1,220 x 102 x 12.7 mm) by gas metal arc welding. Nishida analyzed the experimental results generated by Serotta by using the one-dimensional computer program developed at M.I.T [13]. When initial temperatures of the web and the flange were the same, the final distortion after the specimen cooled down completely was negative (see df of Figure 1). When the initial temperature of the web plate was increased, the final distortion changed to positive. It was found that the final distortion could be reduced to practically zero by selecting a proper temperaturedifferneetial beetween the web and the flange plate.

REFERENCES

[1] Masubuchi, K., *Analysis of Welded Structures*, Pergamon Press, 1980 and a video course with the same title developed by the Center for Advanced Engineering Study, M.I.T., 1990

[2] Masubuchi, K., Luebke, W. H., and Itoh, H., "*Novel Techniques and Their Applications for Measuring Out-of-Plane Distortion of Welded Structures*," Journal of Ship Production, SNAME, Vol. 4, No. 2, May 1988, 73-80

[3] Kolodziejczak, G. C., "*GMAW Control to Minimize Interference from Tack Welds*," Ph. D. Thesis, M.I.T., May 1987

[4] Fujita, Y. et al., *"Studies on Prevention of End Cracking in One-side Automatic Welding,"* Journal of the Society of Naval Architects of Japan, Vol. 136, 1974, 459-465

[5] Chang, I. H., *"Analysis and Control of Root Gap During Butt Welding,"* Ph. D. Thesis, M.I.T., January 1988

[6] Park, S. W., *"In-Process Control of Distortion During Aluminum Butt Welding,"* M.S. Thesis, M.I.T., August 1988

[7] Miyachi, H., *"Control of Thermal Forces Acting on Tack Welds During Butt Welding,"* Ph. D. Thesis, M.I.T., February 198

[8] Bass, R. A., *"Reduction of Residual Stresses and Distortion in HY100 and HY130 High Strength Steels During Welding,"* M.S. Thesis, M.I.T., June 1989

[9] Vitooraporn, C., *"Experimental and Analytical Study on Reduction of Residual Stresses and Distortion During Welding in High Strength Steel,"* Ph. D. Thesis, M.I.T., January 1990

[10] DeBiccari, A., *"A Control of Distortion and Residual Stresses in Girth-Welded Pipes,"* Ph. D. Thesis, M.I.T., July 1986

[11] Barnes, P. K., *"Reduction of Residual Stresses and Distortion in Girth-Welded Pipes,"* M.S. Thesis, M.I.T., tial June 1987

[12] Serotta, M. D., *"Reduction of Distortion in Weldments,"* M. S. Thesis, M.I.T., March 1975

[13] Nishida, M., *"Analytical Prediction of Distortion in "Welded Structures"*, M.S. Thesis, M.I.T., March 1976

Practical Applications of Residual Stress Technology, Conference Proceedings, Indianapolis, Indiana, USA, 15-17 May 1991

Measurement and Prediction of Residual Elastic Strain Distributions in Stationary and Traveling Gas Tungsten Arc Welds

K.W. Mahin and W.S. Winters
Sandia National Laboratories
Livermore, California

T. Holden and J. Root
Atomic Energy of Canada Ltd. Research
Chalk River, Ontario, Canada

ABSTRACT

The residual elastic strain distributions developed in stationary and single-pass gas tungsten arc welds on 304L stainless steel were determined by neutron diffraction and finite element simulation techniques. Comparison of the experimental measurements and the numerical calculations was done on the basis of residual elastic strains, rather than residual stresses, to avoid introduction of error into the experimental data, which measures strain. Using the neutron diffraction technique, reliable experimental measurements of the residual elastic strain distributions were obtained both through-the-thickness and in the plane of the welds. This data provided an excellent experimental basis for comparison with the numerical calculations, allowing us to assess both the accuracy of the analysis and the sensitivity of the analysis to different boundary condition assumptions. The numerical analysis was performed with a fully coupled thermal-mechanical code in which the equations of motion and heat conduction are solved simultaneously and heat transfer due to conduction, radiation, and natural /.forced convection from a free surface is taken into account. The thermal portion of the analysis was tuned using experimentally determined thermocouple histories. However, as the objective of the work was to assess the ability of the deformation model to predict residual elastic strains, hence residual stresses, no tuning was performed in the mechanical portion of the analysis.

The two types of welds studied in this work provided a range of model, as well as experimental complexity. The stationary weld, although easier to model numerically, posed a number of problems experimentally. The reverse was true for the traveling GTA weld. For both welds, the residual elastic strain distributions were measured and computed for the entire welded specimen, not just part of the specimen. In the case of the traveling arc weld, the specimen measured 20 cm x 25 cm (8 inch x 10 inch); for the stationary weld, the specimen was cylindrical, measuring 15 cm (6 inch)in diameter. This paper discusses many of the issues associated with accurately modeling or measuring the residual elastic strains and indicates the conditions under which the numerical calculations were able to accurately predict the residual elastic strain distribution in actual welded specimens.

THE LARGE THERMAL GRADIENTS produced during fusion welding lead to high residual stresses which can significantly affect component performance and reliability by causing deformations or by providing suitable biaxial stress fields which promote crack growth. Although significant progress has been made in computer modeling of the temperature distributions generated during welding [1], quantitative analytical predictions of stresses and strains are still quite limited [1,2]. In practice, detailed measurements of elastic strain fields in welded parts are infeasible from a manufacturing standpoint. However, an experimentally verified computer model which allows calculation of the thermal history and the resultant residual stress distribution is a potentially very important design tool for the engineer. Since the deformation in the material is thermally induced, calculation of this deformation requires an accurate description of both the heat transfer from the heat source and the behavior of the material. Once the heat input into the workpiece has been verified by comparison with experimental measurements of temperature, the residual elastic strains or stresses generated by these thermal fields must be verified. The objective of this work has been to experimentally determine the residual elastic strain fields in a traveling and a stationary GTA weld using neutron diffraction techniques and to compare these measurements to numerical (finite element method) predictions.

EXPERIMENTAL SETUP

TRAVELING GTA WELD - The first weldment studied was an annealed 304L stainless steel plate, 208 mm long, 153 mm wide, and 4.7 mm thick, rigidly clamped in a picture-frame fixture with a full-penetration single-pass autogenous GTA weld placed along the center of the plate, parallel to the long dimension (Figure 1). Experimental determination of the residual elastic strain distribution in the welded plate was performed on the entire welded plate with and without the fixture in place. The average grain size of the annealed material was on the order of 50 μm. The weld

*Work supported under U. S. Department of Energy Contract DE-AC04-76DP0089

was made using 257A, 17.5 V and a speed of 3.3 mm-sec^{-1} with argon gas shielding on the top side. To promote formation of a parallel-sided full penetration weld, small amounts of oxygen (500-750 ppm) were added to the argon shielding gas. The resultant weld was 7.5 mm in width on the top side of the plate, narrowing to a parallel-sided weld 6.0 mm wide at a depth of about 1.0 mm below the top surface. To prevent buckling, the plate was clamped during welding, cool-down and for the initial neutron diffraction measurements. Zircar insulation was used to isolate the specimen thermally from the fixture. Figure 1 shows a schematic of the traveling GTA weld specimen and the corresponding picture-frame clamping fixture used for restraining the specimen during welding and during the initial neutron diffraction analyses. The majority of the neutron diffraction measurements were made along a line perpendicular to the mid-point of the weld. Other detailed measurements were made along a line perpendicular to the "1/4 position" [12].

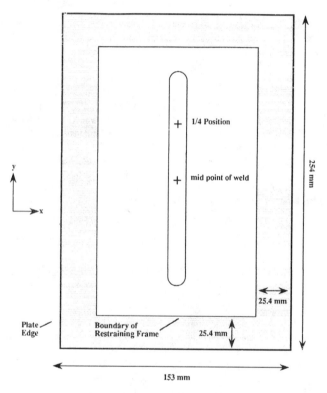

Figure 1. Schematic of the traveling GTA weld specimen and the corresponding picture-frame clamping fixture. The component directions for the measurements and the analysis: y is the weld direction, z is perpendicular to the plate, x is the transverse direction.

AXISYMMETRIC STATIONARY WELD - The second weldment studied was an autogenous axisymmetric stationary partial penetration GTA weld. The workpiece was unrestrained by any external clamping fixture, in constrast to the traveling GTA weld specimen. The weld was located in the center of a circular disk of 304L stainless steel barstock with an average grain size of 150 μm at the edge of the specimen to 100 μm at the center of the specimen. The disk measured 148 mm in diameter with a 2 mm thick, 50 mm high restraining boss around the outer circumference. A cross section of the specimen is shown in Figure 2. The thickness of the welded section was 8 mm. The weld was made at 208A, 14V (avg.) with a dwell time of 5.32 seconds and an argon shielding gas modified with 1000 ppm oxygen to control weld pool shape. The resultant weld measured 4.6 mm in diameter with a penetration depth of 4.5 mm. The residual elastic strains in the specimen were determined using the entire specimen, as welded, i.e. no sectioning of the specimen was done prior to measurement of the strains.

NUMERICAL ANALYSIS

The welding experiments described in the previous sections were modeled with the computer code PASTA2D [3]. PASTA (Program for Application to Stress and Thermal Analysis) is a new finite element code designed to model coupled thermal-mechanical (CTM) behavior in deforming solids. The code is presently configured to solve transient 2-D planar and 3-D axisymmetric problems. The equations of motion and energy for the solid are spatially discretized into four-node quadrilateral finite elements. Spatial integration for the equation of motion is accomplished using one point quadrature with hourglass stabilization and variable artificial bulk viscosity, a technique similar to that utilized in other exclusively mechanical codes such as DYNA2D [4] and PRONTO2D [5]. Spatial integration for the energy equation is accomplished using a full four point quadrature integration over the element. Energy transport within the solid is currently restricted to heat conduction only. The method for modeling the thermal part of the problem is similar to that used in heat conduction codes such as TACO [6]. Time integration of the equations of motion and energy are performed using the explicit time-centered-difference method. Temperature-dependent properties for 304L [7] were included in the calculation. The material behavior of 304L is modeled using the temperature and rate dependent elastic-plastic continuum model proposed by Bammann [8]. A more complete description of PASTA2D, as applied to welding problems, is available in references 2 and 9.

ANALYSIS OF THE TRAVELING GTA WELD - The traveling arc weld problem was modeled as a 2D planar, plane stress problem with the restraining fixture in place (Figure 1). The purpose of generating a parallel-sided full-penetration weld, experimentally, was to provide a quasi-3D problem for the numerical analysis that could be approximated using PASTA2D. Radiative and convective losses from the top and bottom surfaces of the plate were accounted for by using local element volumetric heat loss functions. 3-D thermal transfer effects in the 2-D analysis were accounted for using time-dependent volumetric heat generation functions developed by Kanouff [10], which were constructed using TACO3D [6]. The predicted temperature histories were validated by comparing them with the actual temperature variations measured by thermocouples during and after welding. The mechanical restraint from the fixture was modeled as a "zero-displacement" boundary condition, which, as will be discussed later, lead to discrepancies between the numerical results and the experimental measurements. At the time this problem was modeled, there was no provision in the code for dynamically relaxing, i.e. zeroing, the accumulated strains in the molten region. This caused a significant error in the presentation of the residual elastic strains in the region of the weld pool, but did not impact the accuracy of the strain rate calculations in this region. For the axisymmetric problem, accumulated strains were set to zero in the molten regions and allowed to evolve properly upon solidification and cooldown. Reference 11 contains a more detailed description of the thermal-mechanical analysis.

ANALYSIS OF THE AXISYMMETRIC STATIONARY WELD - For the stationary GTA weld, the physics of the heat source and the molten weld pool were taken into account semi-empirically. The distribution of the GTA heat source on the top surface of the specimen was modeled as a simple Gaussian heat flux distribution with a 2σ radius (defined as the radius over which 98% of the heat is distributed over the surface of the plate) of 2.5 mm. The heat loss due to evaporation in the vicinity of the arc was modeled using a modified Langmuir equation with an efficiency factor of 20% [2]. Since the code computed only heat transfer due to conduction, the heat transfer effects due to fluid flow were approximated by linearly enhancing the thermal conductivity of the liquid. The 2σ radius, the efficiency factor for evaporation, and the enhanced thermal conductivity were considered to be "free" parameters in the analysis and were used to bring the thermal calculations into agreement with experiment. Coarse "tuning" was done by matching the calculated fusion zone contour (1673 K) just prior to arc termination (5.32 sec) to the metallographic cross section of the weld. Additional fine tuning was accomplished by bringing the calculated temperature histories into agreement with the temperature histories measured by twenty-six thermocouples [2]. Once the temperature distributions were verified, a fully coupled thermal-mechanical analysis was performed to determine the residual stress distributions [2]. The weld specimen, including the boss region, was modeled as an unconstrained axisymmetric solid (Figure 2). No tuning was done in the mechanical portion of the analysis. A more detailed description of the thermal and mechanical modeling considerations are provided in references 2 and 9.

from the (113) planes of a Ge single crystal was different for each experiment but was determined precisely in each case by calibrations with standard Si or Ge powders. For the neutron diffraction measurements, the weldment was mounted on a computer-controlled translator table so that any point in the sample could be brought into the gauge volume with a precision of 0.1 mm. At each location, three components of strain were determined for both the (111) and the (002) reflections.

Traveling GTA weld - In the measurement of the transverse and through-thickness components strain in the traveling arc weld, the gauge volume was defined by 2 mm wide slits in absorbing cadmium masks, which were placed before and after the sample. This gauge volume provided good spatial resolution (~ 2 mm x 2 mm) in the x and z directions (Figure 1). In the measurement of the longitudinal, y-component of strain, a 2 mm wide by 25 mm long slit was used. This gave good resolution in the x direction but provided averaged measurements through the thickness of the specimen and along a 25 mm length of the weld. A full description of the experimental setup for this specimen is detailed in references 11 and 12.

Stationary GTA Weld - For the strain measurements in the stationary weld, the slit widths in the absorbing cadmium (Cd) mask were reduced to 1.5 mm and the height was restricted to 2 mm. This produced an approximately cubical gauge volume. Although a small gauge volume was desirable to capture the steep gradients existing around the weld, a considerable amount of scatter appeared in the measurements due to the conflict between the small gauge volume and the relatively large grain size of the material. Discussion of this problem, along with a detailed description of the experiments is provided in references 2 and 13.

DESCRIPTION OF THE TECHNIQUE - When the conditions for constructive interference of neutrons scattered by the atomic lattice are met, a diffraction peak is

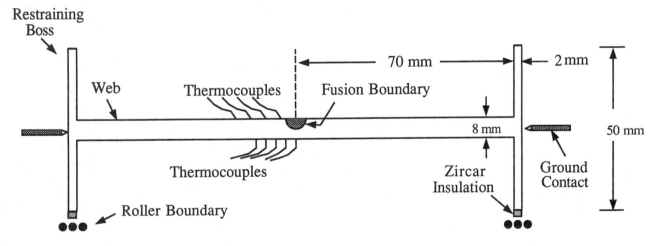

Figure 2. Schematic of the axisymmetric stationary GTA weld specimen, showing the location of the restraining boss and the thermocouple relative to the weld pool region [2,13].

NEUTRON DIFFRACTION ANALYSIS

EXPERIMENTAL SETUP - The neutron diffraction measurements were carried out on spectrometers at the NRU reactor at the Chalk River Laboratories of AECL Research (Atomic Research of Canada). The wavelength of the incident monochromatic beam, provided by reflection

obtained. The interplanar spacing, d_{hkl}, for planes specified by the Miller indices, (hkl), is related to the scattering angle, $2\theta_{hkl}$, by Bragg's law

$$d_{hkl} = \lambda / (2\sin\theta_{hkl}) \quad (1)$$

where λ is the calibrated neutron wavelength. The value of the $2\theta_{hkl}$ is obtained by least squares fitting a Gaussian line-shape on a sloping background to the experimental data of counts versus scattering angle. The precision of the d-spacing measurements for these experiments was about ±0.00015 Å. To determine the strains from the interplanar spacings, a "zero strain" reference sample, extracted from an unwelded, "stress-free" portion of material, was examined for each of the welds analyzed. The error in the strain values was about ±1x10^{-4}. Although the grain structure in the weld is columnar and epitaxial, as compared with the random grain orientation in the base metal, this does not pose a problem for the neutron diffraction measurements, as this texture is directly evident from the intensity plots. Since the as-solidified 304L weld metal is austenitic (fcc) with less than 4 vol.% of ferrite (bcc) phase, similar diffraction conditions can be used in both the weld metal region and the austenitic base metal, which typically has up to 1.5 vol % ferrite in the matrix.

In order to compare the crystallographic strains measured by neutron diffraction with the polycrystalline strains computed in the numerical analysis, it was necessary to select diffraction conditions that represented the maximum and minimum strains in the polycrystal, thus providing a bound for the calculation. For an fcc crystal, such as 304L stainless steel, these are the (111) and (002) reflections, which correspond to the elastically hard ($E_{<111>}$ = 300 GPa) and the elastically soft ($E_{<002>}$ = 100 GPa) directions in 304L stainless steel (E_{bulk} = 189 GPa) [2,14].

NEUTRON DIFFRACTION RESULTS

TRAVELING GTA WELD: The largest tensile component of strain occurred in the longitudinal (ε_y) direction due to shrinkage of the weld metal along the length of the weld. Figures 3a and 3b show the distribution of tensile strains for the (002) and (111) reflections, respectively, as a function of transverse offset (x) from the weld center. Note that the maximum longitudinal strain (ε_y) does not appear in the weld metal (± 3 mm wide) but outside the weld about 9 mm from the weld center. As expected from the difference in the elastic constants, the longitudinal strains for the (002) reflections are larger than for the (111) reflection (Figures 3a and 3b), i.e. more deformation has occurred with the lower elastic modulus. In contrast to the longitudinal strains, the transverse (ε_x) and through-thickness (ε_z) components of strain are much smaller and are compressive with typical maximum values on the order of -6x10^{-4} and -2x10^{-4} respectively for the (002) reflection and in the same range for the (111) reflection. A more detailed discussion of the measurements is provided in references 11 and 12.

One feature which is not shown in Figures 3a and 3b is the very strong preferred orientation that exists in the weld metal, as a result of selective epitaxial growth [11,12]. It is found, for example, that the number of (002) oriented grains is an order of magnitude greater than the number of (111) grains for the longitudinal and transverse directions [12]. This means that in the weld metal the (002) strains are a much more representative measure of the residual elastic strains than the (111) strains.

STATIONARY GTA WELD: Figures 4a-c show the measurement (symbols) of the variation in residual elastic strain, as a function of distance from the weld centerline, for the (002) and (111) reflections obtained from the partial penetration weld at the mid-thickness of the disk (4.0 mm). The mid-plane intersects the very bottom of the weld nugget, which has a penetration depth of 4.5 mm. The considerable scatter in the neutron diffraction data for this weld, in comparison with the traveling arc weld experiment, is associated with the inability of the small gage volume to encompass a sufficiently large enough sample of properly oriented grains to obtain a statistical sample of the polycrystalline, rather than the grain-to-grain, strain data. The through-thickness component of strain for the axisymmetric weld is strongly compressive (Figure 4a), although it might be anticipated in an 8 mm thick plate that the through-thickness stresses would be quite small. The radial and hoop strains (Figures 4b and 4c, respectively) are positive for the stationary weld and have the same magnitude in the as-solidified weld pool region as would be expected from symmetry. The radial strains fall off more slowly with radial offset than do the hoop strains. The solid curves are

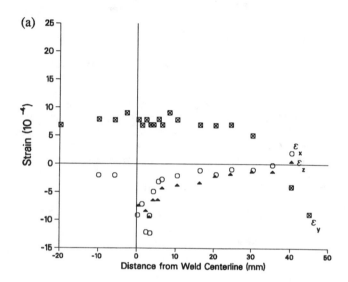

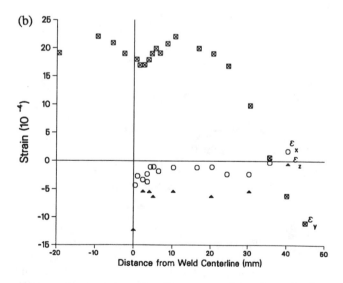

Figure 3. Plots of the neutron diffraction measurements for the longitudinal, ε_x, the transverse, ε_y, and the through-thickness, ε_z, components of residual elastic strain in the traveling GTA weld specimen. (a) <111> reflection and (b) <002> reflection.

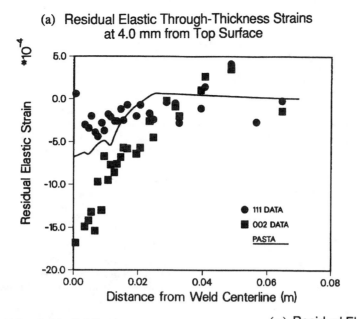
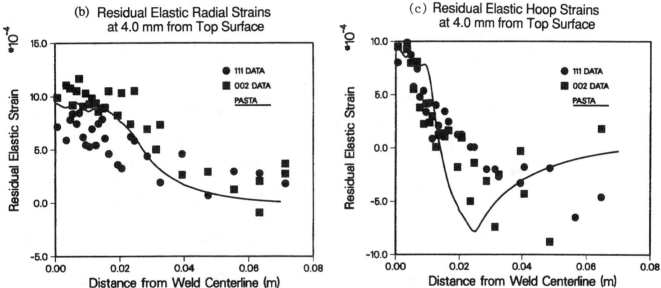

Figure 4. Comparisons between the residual elastic strain distributions determined for the stationary GTA weld specimen as a function of location and direction. The symbols correspond to the neutron diffraction measurements taken for the <111> and the <002> reflections; the straight lines represent the PASTA calculations. (a) through-thickness strains, (b) radial strains, and (c) hoop strains.

the calculated elastic strain distributions generated from the numerical analysis. In the analysis, the strains are derived from the stresses using the bulk elastic constants for an isotropic material [2,9].

COMPARISON OF NUMERICAL PREDICTIONS WITH EXPERIMENTAL MEASUREMENTS

Verification of the code predictions was based on comparison of the residual elastic strains, rather than on comparison of the residual stresses, although the code calculates both. An assessment of the residual stress in general requires measurements of all components of the strain tensor and a knowledge of the elastic constants. In the case of the neutron diffraction data, it was assumed that the components of strain measured for the (111) and (002) directions, i.e. the longitudinal, transverse, and through-thickness directions for the traveling weld and the radial, hoop, and through-thickness directions for the stationary weld, coincided with the principal strain axes. In order to avoid the introduction of error by trying to convert the measured strains to stresses, the residual stress distributions calculated in PASTA2D were verified by comparing the measured strains to the calculated residual elastic strains. As mentioned previously, the (002) and (111) diffraction conditions were selected to provide an upper and lower bound on the polycrystalline PASTA2D strain calculations. The most accurate predictions fall between the two sets ((111) and (002)) of neutron diffraction data.

TRAVELING GTA WELD: Figure 5a shows the longitudinal (ε_{yy}) residual elastic strains calculated in PASTA2D (solid line) plotted against the neutron diffraction measurements (symbols). Between 0.8 cm and 3.0 cm from the weld centerline, the strain data, calculated in PASTA2D assuming isotropic properties, falls between the two experimental curves, indicating a good correlation between the actual and the predicted residual elastic strains. At distances greater than 3.0 cm, the effect of the discrepancy between the assumed behavior for the fixture (zero-displacement) and the actual behavior (some slippage or movement) causes disagreement between the numerical

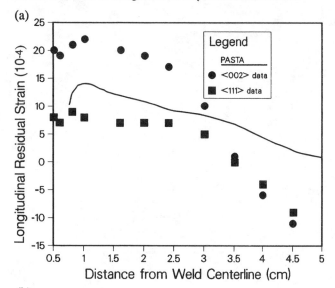

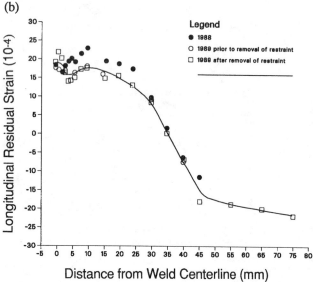

Figure 5. (a) Comparison between the longitudinal strains measured, as a function of position, by neutron diffraction (symbols) and those calculated in the PASTA analysis (solid line) for the traveling GTA weld specimen; (b) longitudinal elastic strains measured by neutron diffraction in the welded specimen with and without the restraint in place.

predictions and the experimental measurements. Figure 5b shows the neutron diffraction measurements taken after one year with the fixture removed. In both cases, the residual elastic strains at the boundary were compressive, rather than tensile, as calculated with the zero displacement boundary condition. At distances less than 0.8 cm, the inability of the code (at the time the calculation was initially performed) to "zero" the accumulated strains at melt, caused a significant deviation between the calculated strains and the measured strains. As a result of these discrepancies, it was notpossible with this experiment to properly assess the effect of the 2-D assumption on the accuracy of the numerical strain predictions.

STATIONARY GTA WELD: Figures 4a-c show the comparison between the strains calculated in PASTA2D (solid) line and the neutron diffraction measurements (symbols). Unlike the case of the traveling arc weld, the "restraining fixture" - in this case the restraining ring machined around the outside of the specimen - was included in the model and "slippage" between the part and the fixture was not an issue. Similarly, PASTA2D was modified for this calculation to allow "zeroing" or dynamic relaxation of the accumulated strains at melt. As a result of the modifications made to the experimental design and to the model, the agreement between the PASTA2D results and the experimental measurements was excellent for all three components of strain at the mid-thickness of the plate (Figures 4a-c). Comparisons between the calculations and the measurements for four other depth locations through the thickness showed similar agreement [2].

CONCLUSIONS

The residual elastic strains calculated in the "polycrystalline" PASTA2D analysis agreed well with the residual strains measured on the (111) and the (002) crystallographic planes for the axisymmetric stationary GTA weld experiment. The discrepancies noted in the traveling arc weld experiment are believed to be due to the failure of the calculation to account for movement of the specimen in the fixture and for the zeroing of the accumulated strains at melt. As a result of these two calculations and other data, we have concluded that the temperature and strain rate dependent material model used for 304L stainless steel will accurately predict the residual elastic strain and hence stress distribution in a GTA weld, as long as the thermal solution is successfully "tuned" to reflect the energy deposition into the workpiece. In both the traveling and the stationary GTA weld experiments, the thermal "tuning" was accomplished by matching the temperature histories recorded by at least eight thermocouples with the corresponding temperature histories calculated for similar locations within the simulation.

REFERENCES

1. Goldak, J. A., "Modeling Thermal Stresses and Distortions in Welds", Proceedings of the 2nd Int'l Conf. on Trends in Welding Research, ed. S. A. David, publ. ASM Int'l, 71-82, (1989)
2. Mahin, K. W., W.S. Winters, T. M. Holden, R. R. Hosbons, S. R. MacEwen, "Prediction and Measurement of Residual Elastic Strain Distributions in Gas Tungsten Arc Welds", accepted for publication in the Weld. Jnl (1991)
3. Winters, W. S. and W. E. Mason, "PASTA2D-Applications," SAND89-8217, Sandia National Laboratories, Livermore, Ca (1989)
4. Hallquist, J. O., "User's Manual for DYNA2D - An Explicit Two-Dimensional Hydrodynamic Finite Element Code with Interactive Rezoning," UCID-18756, University of California, Lawrence Livermore National Laboratory, Livermore, Ca (1980)
5. Taylor, L. M. and D. Flanagan, "PRONTO2D, A Two-Dimensional Transient Solid Dynamics Program," SAND86-0594, Sandia National Labs, Albuquerque, NM (1987)

6. Mason, W. E., "TACO3D - A Three-Dimensional Finite Element Heat Transfer Code," SAND83-8212, Sandia National Laboratories, Livermore, Ca (1983)

7. Leibowitz, L., "Properties for LMFBR Safety Analysis" ANL-CEN-RSD-76-1. 116 pp. (1976)

8. Bammann, D. J., "Parameter Determination for a Strain Rate and Temperature Dependent Plasticity Model," SAND89-8812, Sandia National Laboratories, Livermore, CA (1989)

9. Winters, W. S. and K. W. Mahin, "PASTA2D Modeling of an Axisymmetric Stationary Gas Tungsten Arc Weld", Modeling of Casting, Welding and Advanced Solidification Processes, proceedings of the Engineering Foundation Conference in Davos Switzerland, Sept. 1990, ed. M. Rappaz, M. Ozgu, K. W. Mahin, publ. TMS-AIME (1991)

10. Kanouff, M. P. and W. M. Mason, "Weld Modeling Using a Generalized Three Dimensional Heat Conduction Code", published in the Abstract Book for the 69th Annual AWS Convention, New Orleans (1988)

11. Mahin, K. W., S. MacEwen, W. S. Winters, W. E. Mason, M. Kanouff, E. A. Fuchs, "Evaluation of Residual Stress Distributions in a Traveling GTA Weld Using Finite Element and Experimental Techniques" Modeling and Control of Casting and Welding Processes IV, ed. A. F. Giamei, publ. TMS-AIME, 339-350 (1988)

12. Holden, T. M., R. R. Hosbons, J. H. Root, S. R. MacEwen, "Residual Stresses in a Single Pass Weld in a 304L Stainless Steel Plate" ANDI-31, Chalk River Nuclear Labs, Canada (1989)

13. Holden, T. M., R. R. Hosbons, S. R. MacEwen, D. V. Parsons, "Neutron Diffraction Measurements of the Distribution of Residual Strains in and near Stationary Gas Tungsten Arc Welds in 304 Stainless Steel" ANDI-32, Chalk River Nuclear Labs, Canada (1989)

14. Kikuchi, M., "Elastic Anisotropy and Its Temperature Dependence of Single Crystals and Polycrystal of 18-12 Type Stainless Steel" Trans. of Jpn. Inst. of Metals, 12, 417-421 (1971)

Residual Stress Measurements at Skip Fillet Welds

W.Y. Shen, P. Clayton, and M. Scholl
Oregon Graduate Institute
Beaverton, Oregon

ABSTRACT

Fillet welds used to attach stiffeners and brackets to railroad tank car shells generate residual stresses. As part of a program to investigate the effect of as-welded residual stresses on the fatigue life of the shell material, stress patterns have been determined experimentally. The information has been used to evaluate the validity of previous theoretical calculations, develop a fatigue specimen to simulate the full scale welded structure and establish the effects of cyclic loading on residual stress.

Several 19 x 2½ x ½ in. pads of A283 steel were skip fillet welded to 27 x 10½ x ½ in. A515 steel baseplates with a two-pass shielded metal arc welding procedure in compliance with industry standards. The residual stress patterns in the base plate were constructed from measurements made on different specimens using both blind hole drilling and sectioning methods. The stresses were biaxial tension in the material adjacent to the center of the weld and biaxial tension-compression at the weld ends.

Smaller specimens, suitable for fatigue testing, were made using bead-on-plate and fillet welds. For specimens greater than 4 in. width the residual stress patterns were similar to those in the pad-on-plate specimens. After longitudinal, pulsed tension, cyclic loading with a load range greater than ⅓ the yield strength, 15 Ksi, the longitudinal residual stresses were significantly relieved.

INTRODUCTION

The potential problem of welding residual stress on the fatigue life of tank car shells has recently been addressed by Tong et al [1] and Orringer et al [2]. They considered some older railroad tank cars which were strengthened by the addition of under-belly spanning stiffeners. The stiffener was continuously fillet welded to narrow rectangular pads which were skip fillet welded to the tank car shell. The weld pattern and pad dimensions are shown in Figure 1.

The concern is that a combination of live stress and residual stress could at some time in the future result in the development of through cracks in the tank car shell. Although live stresses have not been determined for the pad-shell weld locations for these vehicles it is considered that they will involve low, longitudinal, tension-to-tension fluctuating stresses resulting from flexing of the vehicle, and low longitudinal tension-to-compression stresses caused by acceleration and deceleration of the vehicle. This applied stress field would make the weld toe at a skip weld end the most likely site of fatigue crack initiation.

The combination of high mean stress (welding residual stress) and very low amplitude live stress is one that tends to fall outside the normal ranges of laboratory fatigue data. Gurney and Maddox [3-5] have carried out the most comprehensive studies of the effect of residual stress on the fatigue behavior of fillet welded specimens. Their conclusion was that for an as-welded specimen with thickness less than 13mm there is no significant effect of residual stress on the fatigue life if the applied loading is under tension-tension conditions. However, they have few data points at very low applied stresses and have not specifically studied weld ends. Thus, results for the effect of welding residual stresses on the fatigue behavior of welded structures are still not conclusive [6-13].

Theoretical calculations for skip welds indicate [1-2] that there are biaxial tension-tension residual stresses at the weld end with equivalent stresses of yield level within a critical zone about ¼ in. in diameter. Although it is commonly accepted [6-14] that very high tensile residual stresses do occur, there are only limited experimental data to define a two dimensional welding residual stress pattern at a skip weld end.

One objective of the present work was to use strain gages to measure the residual stress patterns associated with skip fillet weld ends for simulated tank car pad welds, made to the same specifications and practices used by the manufacturer. The practical limitation of strain gage size made it necessary to construct such a pattern from the composite data for many different welds. The measurements were compared with theoretical calculations to assess the validity of the model used.

A second goal was to develop a smaller, and simpler, test specimen with which to determine the effect of residual stress on the fatigue life of A515 grade 70 steel. It was important to have the same residual stress pattern as in the more complex pad-on-plate situation but for testing purposes the specimen needed to be as compact as possible. This necessitated determining the effect of specimen size on the residual stress pattern.

The test specimen that evolved was used to determine the effect of cyclic loading on the level of residual stress. The purpose was to define the values of applied stress which result in residual stress relaxation. Subsequent work will be carried out to determine the effect of residual stress on the fatigue life of A515 at applied stresses which do not cause residual stress relaxation.

EXPERIMENTAL PROCEDURE

Materials: ASTM A515 Grade 70, ½ in. steel plate and ASTM A283 ½ by 2½ in. steel strap were used to fabricate the test specimens. The chemical composition of the steels used are shown in Table I. The A515 plate had a 45 ksi yield strength and a 70 ksi ultimate tensile strength with 37.5% elongation in 2 inches and 60% reduction-in-area.

Welding Procedure: Welding of the specimens was carried out following the relevant industrial specifications by a qualified welder using the shielded metal arc welding process. Low-hydrogen E7018 electrodes [15], 3/16 in. diameter, were used at an arc current of 200 amps with direct current reverse polarity. Fillet welds were made in two passes with a weld throat not exceeding ¼ in.

Pad-on-Plate Specimen: Full scale specimens to simulate the pad-tank shell structure were fabricated using an A515 base plate and an A283 pad. Dimensions and weld pass sequence are shown in Figure 2.

Simple Specimen: The simple specimen, Figure 3, was developed in conjunction with the residual stress measurements. Two variations were employed; one with a bead-on-plate weld, and the other with a 3 x ½ x ½ in. bar welded to the specimen by a single sided fillet weld. Welding parameters were identical to those used on full-scale specimens except for the travel speed which was decreased for the bead-on-plate welds.

Residual Stress Measurements: Both sectioning and hole drilling techniques were used to measure the residual stresses. Measurements were made predominantly at the weld ends and weld center. For sectioning measurements, 4.3 x 4.8 mm three element strain gage rosettes (MicroMeasurements WA-06-030WR-120) were used with a gap of 4 mm between neighboring gages. Sections were cut using a 0.5 mm wide cold saw. Figure 4 depicts the measurement sites while Figure 5 shows the coordinate system used to define location.

For residual stress measurements by hole drilling 9.6 x 12.2 mm, three element gages, (MicroMeasurements CEA-06-062UM-120) were used. These gages allowed the hole to be drilled as close as 1.5 mm to the welds. Blind holes, 1.5 mm in diameter and 2 mm deep, were drilled using a carbide cutter in a high speed air turbine fixture.

All strain gage measurements were made using a four terminal bridge and a digital multimeter with a resolution of 0.001 ohms. This resolution was equivalent to a strain of 4×10^{-6} in/in. To minimize thermal effects, temperature compensation gages were used for all measurements.

Effect of Specimen Width: The effect of specimen width was investigated using the simple bead-on-plate and single fillet welded specimens. Welding residual stresses were measured using the hole drilling technique on specimens 2, 4, 5 and 8 inches wide.

Cyclic Load Tests: Simple specimens were used to examine the effect of cyclic loading on the residual stresses. Specimens, shown in Figure 3, were tested under pulsating tension in a 200,000 lb. load frame. Tests were performed at load ranges of 6 ksi (41 MPa), 14 ksi (96 Mpa), 22 ksi (150 Mpa), and 30 ksi (205 MPa). The effect of the number of cycles was also investigated over these load ranges at 150, 2000, and 115000 cycles.

RESULTS

Residual Stress Measurements: Results were obtained from several specimens of each type, full-scale, bead-on-plate simple specimens, and fillet-welded simple specimens. Both hole drilling and sectioning techniques were used on the full scale specimens, while hole drilling alone was used on both types of simple specimens. All the data are shown in Table II.

The results obtained from the simple specimens are, in general, in good agreement with those from the pad on plate samples. Consider the x=0, y=0mm coordinate for which there are six data sets in Table II; H6-f, H7-s, hb3-s, hg10-f, hg12-f, and hg13-s. These produce an average of 16.3 ksi for σ_x and -13.8 Ksi for σ_y. The ranges are 8.2 to 22.1 for σ_x and +1.7 to -19.8 for σ_y.

The residual stresses associated with the start end of a weld were very similar to those at the finish end.

Figure 6 represents the overall composite picture and contains all the data from Table II for all types of specimen. For those locations with multiple measurements the results have been averaged. The magnitude of the residual stress is indicated by the length of the line while the arrows differentiate between tension and compression.

Although it is not easy to determine patterns from this complex plot it is possible to see that close to the weld end σ_x is tensile and σ_y compressive while at the weld center the stresses are bi-axial tension.

The pattern of stresses between two weld ends is brought out more clearly for all measurements made at y=0 and 0.5mm in Figure 7. Both σ_x and σ_y decrease rapidly with distance from the weld end. In this case there is excellent correlation between the different specimen types.

<u>Width Effects</u>: Simple specimens 2, 4, 5, and 8 inches wide were used to measure the effect of specimen width on the welding residual stresses. A common measurement point, (x,y = 0,0) was used for each specimen and the resulting σ_x and σ_y values are shown plotted in Figure 8. The residual stresses are essentially constant at a specimen width of 4 inches or greater. On the basis of these results all simple specimens were made 5 in. in width, Figure 3.

<u>Cyclic Load Effects</u>: Simple specimens were used to examine the effect of pulsating-tension applied loads on the welding residual stresses. Figures 9 and 10 show the effects of number of cycles and stress range on σ_x and σ_y respectively. Identical measurement points (x,y = 0,0) were used for all the specimens; Table III shows the results. At load ranges above approximately ⅓ the yield strength, 15 ksi, the magnitude of σ_x decreases quite rapidly while there is little change in σ_y. The residual stress relaxation occurs in the first few hundred cycles.

DISCUSSION

One of the principal aims of this study was to produce experimental data to help validate an earlier theoretical model [1,2]. The model attempts to evaluate the fillet welding of stiffener pads to the shell of a railroad tank car. The concern was that the residual stresses induced by welding would be of sufficient magnitude to compromise the fatigue life of the tank car shell.

The model calculates the welding residual stresses by considering the weld to be anchored at the ends and all stresses applied to the base plate through that end point. The weld was considered to have no width. The calculations reveal tensile welding residual stresses of almost yield magnitude for σ_x within a region ¼ in. in diameter from the weld end point. The value of σ_x decreased rapidly with distance from the weld. The experimental results also indicate that σ_x at the weld end would be close to yield with a steep stress gradient. The major difference between the calculated and experimental data is that, in the former, σ_y close to the weld end was tensile whereas experimental results showed a compressive stress.

Previous studies have shown that the presence of welding defects are influential in shortening fatigue life [16,17]. In the case of the tank car it is likely that a defect at the weld toe of the weld end would be the initiation site of a fatigue crack. Such a defect would be oriented at right angles to the tensile residual stresses and the axis of the applied stress field.

The effect of the specimen width on the welding residual stresses was investigated so that a specimen could be designed for subsequent cyclic loading and fatigue tests. A specimen 5 inches wide has a residual stress pattern similar to that of the full scale plate. Simple specimens were cycled under pulsating-tension at several load ranges and for several cycle durations. Above a maximum stress of about ⅓ the yield, the magnitude of the longitudinal residual stresses decrease rapidly after only a few hundred cycles. At high levels of loading, about 75% of yield, they were almost completely eradicated. The transverse residual stress (normal to the weld axis and loading direction) was not affected in the same way. This behavior is similar to that reported by other investigators [8,18].

The relaxation of welding residual stresses provides a basis for the explanation of results of fatigue tests at high load ranges under pulsating-tension where little if any effect of residual stresses on fatigue life has been found. The effect of welding residual stresses on fatigue life at low-stress ranges is less clear and will form the basis of a future fatigue investigation.

CONCLUSIONS

1. Consistent residual stress patterns were observed at the ends of fillet welds and bead-on-plate welds. The stresses were tensile in the weld longitudinal direction and compressive in the transverse direction.

2. A significant relaxation of the longitudinal welding residual stress at the weld end resulted from cycling under pulsating tension with a maximum applied load of one third the yield strength of the steel.

ACKNOWLEDGEMENTS

The authors wish to thank the Federal Railroad Administration Office of Research and Development for funding the work and especially Dr. O. Orringer, U.S. Department of Transportation, Volpe National Transportation Systems Center, for many helpful discussions. They also thank Bob Turpin for making the weld specimens and the General American Transportation Corporation (GATX) for providing the welding procedure specifications and other useful information.

REFERENCES

1. Tong, P. et al, "DOT-111A/100W tank cars special retrofit stiffener integrity assessment", task force report, U.S. DOT Transportation Systems Center, Cambridge, MA, 1987.

2. Orringer, O., J.E. Gordon, Y.A. Tange and A.B. Perlman, "On some problems of stress concentration in railroad tank car shells", App. Mech. Rail Transportation Symposium - 1988, ed. by H.H. Barr, T. Herbert, and M. Yovanovich, ASTM, 1988, pp. 88-94.

3. Gurney, T.R. "Some recent work relating to the influence of residual stresses on fatigue strength", Residual Stress in Welded Construction and Their Effect, Welding Institute Conference, London, 15-17 Nov. 1977.

4. Maddox, S.J., "Influence of tensile residual stress on the fatigue behavior of welded joints in steel", Residual Stress Effects in Fatigue, STP 776, pub. by ASTM, 1982, pp.63-96.

5. Maddox, S.J., "Fatigue of stress-relieved fillet welds under part-compressive loading", Research Report, Welding Institute, Nov. 1982.

6. Munse, W.H., "Fatigue of welded steel structures", Welding Research Council, New York, 1964.

7. Masubuchi, Analysis of Welded Steel Structures, Oxford, New York, 1980.

8. Berge, S. and O.I. Eide "Residual stress and stress interaction in fatigue testing of welded joints", Residual Stress Effects in Fatigue, STP 776, pub. by ASTM, 1982, pp. 115-131.

9. Harrison, J.D., "The effect of residual stresses on fatigue behaviour", Residual Stresses and Their Effect, pub. by The Welding Institute, 1981, pp. 9-16.

10. Fukuda, S. and Y. Tsuruta, "An experimental study of redistribution of welding residual stress with fatigue crack extension", Transactions of JWRI, 7 (2), 1978, pp. 67-72.

11. Fukuda, S., S. Watari, and K. Horikawa, "An experimental study of effect of welding residual stress upon fatigue crack propagation based on observation of crack opening and closure", Transactions of JWRI, 8 (2), 1979, pp. 105-111.

12. Itoh, I.Z., S. Suruga, and H. Kashiwaya, "Prediction of fatigue crack growth rate in welding residual stress field", Engineering Fracture Mechanics, 33 (3), 1989, pp. 397-407.

13. Glinka, G., "Effect of residual stresses on fatigue crack growth in steel weldments under constant and variable amplitude loads", Fracture Mechanics, STP 677, C.W. Smith ed., pub. by ASTM, 1979, pp. 198-214.

14. Chien, C.-H., H.-J. Chen, and Y.-T. Chiou, "The estimation of welding residual stresses by using simulated inherent strains", Transactions of the Japan Welding Society, 20 (1), 1989, pp. 52-59.

15. Linnert, G.E., Welding Metallurgy, pub. by AWS, New York, 1967 pp. 73.

16. Engesvik, K.M. and T. Moan, "Probabilistic analysis of the uncertainty in the fatigue capacity of welded joints", Engineering Fracture Mechanics, 18 (4), 1983, pp. 743-762.

17. Bell, R., O. Vosikovsky, and S.A. Bain, "The significance of weld toe undercuts in the fatigue of steel plate t-joints", International Journal of Fatigue, 11 (1), 1989, pp. 3-11.

18. Lu, J., J.F. Flavenot, and A. Turbat, "Prediction of Residual stress relaxation during fatigue", Mechanical Relaxation of Residual Stresses, STP 993, L. Mordfin ed., pub. by ASTM, 1988, pp. 75-90.

TABLE I. Composition of A515 and A283 Steels (in wt.%)

Steel	C	Mn	P	S	Si
A515-70	0.31	0.90	0.035	0.04	0.23
A283	0.18-0.24	0.90	---	---	0.10-0.30

TABLE II. Residual Stress Data

Measurement	x,mm	y,mm	σ_x, ksi	σ_y, ksi
S3-f	-5.3	-2.5	6.7	-20.6
S4-f	-13.8	6.0	-8.5	-26.5
S5-f	-5.3	6.0	-5.4	-15.8
S6-f	3.3	6.0	5.4	0.4
S7-f	11.8	6.0	-5.2	-5.0
S8-f	-13.8	16.0	-23.4	-42.4
S9-f	-5.3	16.0	-16.2	-24.8
S10-f	3.3	16.0	-11.6	-8.4
S11-f	11.8	16.0	-11.8	3.2
S12-f	-13.8	62.0	-14.1	-7.9
H1	-20.0	0.0	3.3	-37.2
H2-f	-12.0	0.0	5.6	-33.5
H3-s	-10.0	0.0	1.4	-32.5
H4-s	-5.0	0.0	5.6	-31.5
H5-f	-5.0	0.0	14.9	-17.7
H6-f	0.0	0.0	16.4	-9.8
H7-s	0.0	0.0	19.5	-18.0
H8-s	-21.0	3.0	-6.1	-47.8
H9-s	-6.0	3.0	2.4	-28.1
H10-f	-3.0	3.0	3.5	-25.7
H11-s	-3.0	3.0	6.9	-16.2
H12-f	-0.5	3.0	20.3	-15.8
H13-s	0.5	3.0	17.7	11.6
H14-s	1.0	2.0	3.0	-7.4
H15-f	1.5	3.0	10.6	0.5
H16-s	2.5	2.0	11.2	-16.5
H17-f	7.0	2.5	8.0	-14.9
H18-c	20.0	2.0	27.0	22.5
H19-c	20.0	3.0	11.8	20.2
H20-c	20.0	3.0	12.9	23.1
H21-f	-11.8	6.0	-1.9	-34.0
H22-s	-7.0	6.0	10.2	-27.9
H23-f	-5.3	6.0	1.5	-11.5
H24-s	-3.0	5.0	3.9	-11.6
H25-s	0.0	6.0	16.3	5.9
H26-f	2.5	6.0	10.4	2.6
H27-s	5.0	6.0	-7.1	-1.1
H28-f	12.0	6.0	4.2	3.6
H29-f	13.5	6.0	8.3	22.9
H30-c	20.0	6.0	7.3	-21.9
H31-f	10.0	9.0	-4.3	-0.6
H32-c	20.0	9.0	-6.2	16.8
H33-c	20.0	9.0	3.1	19.2
H34-c	20.0	11.0	-17.5	17.6
H35-c	20.0	11.0	-24.0	10.3
H36-c	20.0	12.0	-15.8	18.4
H37-f	-11.8	16.0	0.1	-21.9
H38-f	15.0	16.0	-11.7	9.1
H39-f	-9.0	65.0	-16.2	-8.4
H40	20.0	66.0	-20.4	-3.5
H41-f	-8.0	67.0	-18.2	3.7
H42	20.0	67.0	-20.7	-11.1
hb1-f	-3.0	-7.0	23.3	-26.3
hb2-f	-3.0	-4.0	16.5	-26.2
hb3-s	0.0	0.0	22.1	-18.6
hb4-s	0.0	0.5	21.1	-13.6
hb5-s	0.0	0.5	18.6	-10.8
hb6-f	0.0	0.5	16.6	-18.5
hb7-s	5.0	3.0	26.5	-5.1
hb8-c	20.0	3.0	37.8	6.9
hb9-c	20.0	3.0	44.2	2.3
hg10-f	0.0	0.0	19.9	1.7
hg11-c	20.0	3.0	25.0	6.3
hg12-f	0.0	0.0	11.8	-13.6
hg13-s	0.0	0.0	8.2	-14.2

S = sectioning measurements.
H = hole drilling measurement on a full scale specimen.
hb = hole drilling measurement on a simple bead-on-plate specimen.
hg = hole drilling measurement on a simple fillet weld specimen.

f and s are at the finish & start end of welds; c is at the center of a weld.

TABLE III.

Applied Stress Range, ksi	Number of cycles	σ_x, ksi	σ_y, ksi
6	115,000	15.8	-13.8
6	115000	20.6	-7.7
14	2000	10.9	-4.5
14	2000	12.6	0.1
14	115000	15.2	-9.3
14	115000	21.2	-5.5
22	150	1.9	-8.9
22	150	2.8	-20.5
22	2000	1.3	-6.6
22	2000	3.3	-2.2
22	2000	6.2	-13.6
22	2000	2.2	-14.5
22	115000	7.8	-14.6
22	115000	8.2	-23.1
22	115000	3.8	-14.3
22	115000	5.8	-16.8
30	150	-0.3	-7.5
30	150	3.7	-9.2
30	115000	-4.2	-9.7
30	115000	2.6	-3.7
30	115000	-4.3	-11.7

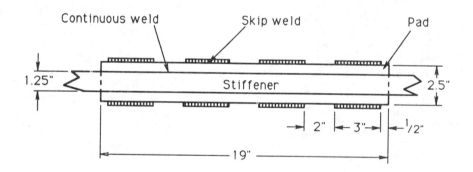

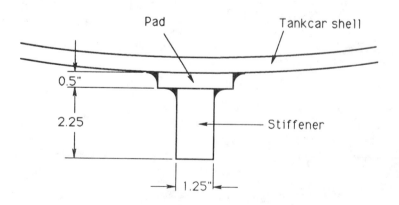

FIGURE 1. Tank car shell/pad/stiffener arrangement.

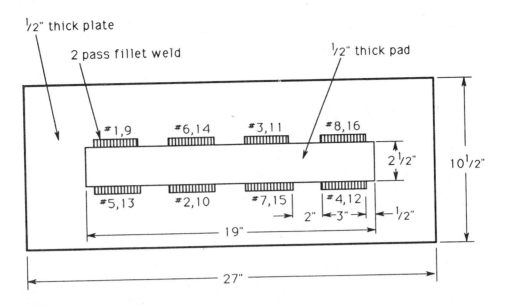

FIGURE 2. Full-scale, pad-on-plate specimen with typical weld pass sequence.

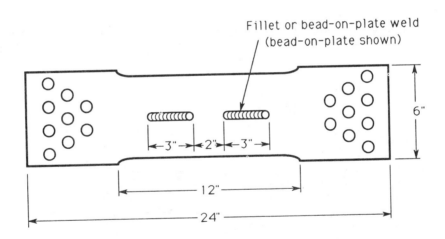

FIGURE 3. Simple specimen configuration and dimensions.

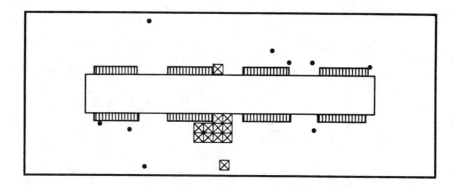

⊠ = Sectioning measurement
• = Hole drilling measurement

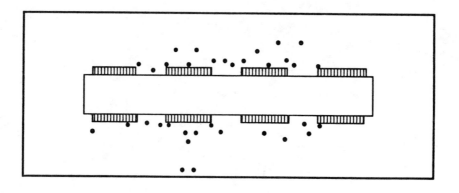

FIGURE 4. Residual stress measurement locations on full-scale specimen.

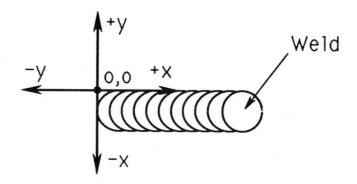

FIGURE 5. Coordinate system used for residual stress measurement location.

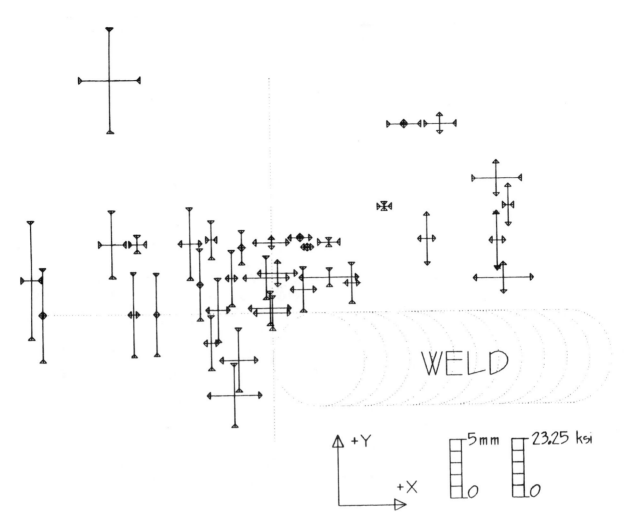

FIGURE 6. Magnitude, direction, and location of measured residual stresses. Line length indicates stress level, inward pointing arrows represent compression, outward pointing arrows represent a tensile stress.

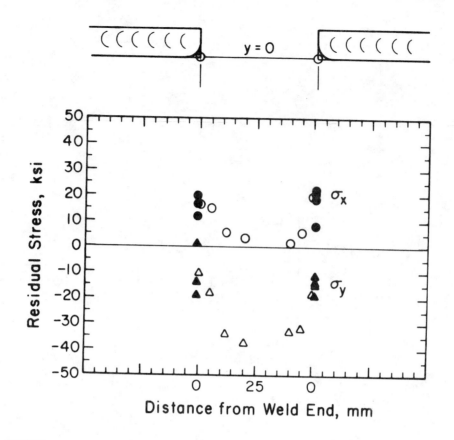

FIGURE 7. Residual stresses between welds at y = 0. Solid markers represent sectioning measurements, open markers represent hole-drilling measurements.

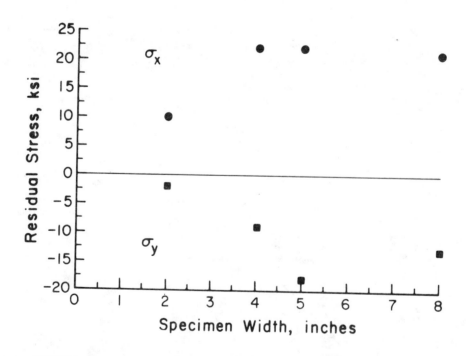

FIGURE 8. Effect of simple specimen width on residual stresses.

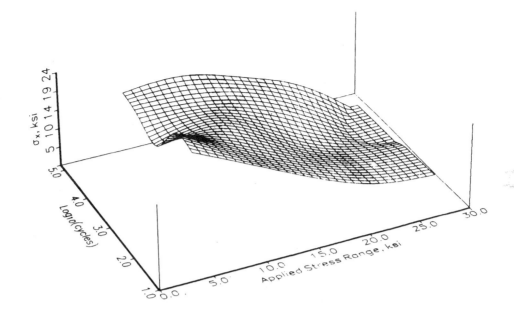

FIGURE 9. Effect of number of cycles and pulsating stress range on residual stress, σ_x.

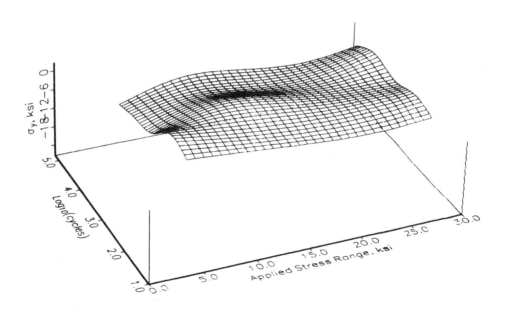

FIGURE 10. Effect of number of cycles and pulsating stress range on residual stress, σ_y.

… Practical Applications of Residual Stress Technology, Conference Proceedings, Indianapolis, Indiana, USA, 15-17 May 1991

Characterization of Residual Stresses in an Eccentric Swage Autofrettaged Thick-Walled Steel Cylinder

S.L. Lee, G.P. O'Hara, V. Olmstead, and G. Capsimalis
Benet Laboratories
U.S. Army Armament Research,
Development, and Engineering Center
Watervliet, New York

ABSTRACT

In the swage autofrettage process, partial autofrettage in the cylinder was achieved by driving an oversized mandrel through the cylinders to cause plastic deformation. In this work, we made experimental and theoretical investigations of residual stresses in a swaged asymmetric thick-walled cylinder with small wall variations. Hoop and radial residual stress distributions, as well as angular hoop stress variations at the inside and outside diameter (ID and OD) of the cylinder, were determined by using a single-exposure position-sensitive scintillation x-ray diffraction detection system. Our experimental results were in good agreement with the residual stress distributions predicted from Tresca's hydraulic autofrettage model. Deviations of hoop residual stress at the bore may be attributed to reverse yielding effects. Finite element analysis using ABAQUS code on a Convex supercomputer was used to model the swage autofrettage process. Small angular residual stress variations were predicted due to the eccentricity of the cylinder. The finite element model predictions of residual stress distribution were in general agreement with experimental results except near the bore. Preliminary results of an alternative finite element method showed a reduced compressive hoop stress near the bore, similar to the results obtained by assuming Bauschinger's effect.

THE AUTOFRETTAGE PROCESS INVOLVES THE application of radial forces at the bore of a cylinder with sufficient magnitude to cause permanent bore expansion. Mechanical swage autofrettage eliminates the ultrahigh pressures required in the conventional hydraulic method to produce the same radial forces [1,2]. As a result, the residual stress distribution increases the elastic strength of the cylinder, retards the growth of fatigue cracks at the cylinder bore, and improves the roundness and straightness of the cylinder. Work on mandrel geometry design requiring minimum load has been reported [3]. However, limited work has been done in the evaluation of the mandrel design and the swage process by residual stress measurements. Furthermore, in the manufacturing process, an asymmetric cylinder resulting from the boring process is generally rejected without knowledge of the effects on residual stresses. In the present investigation, residual stress measurements were made using the x-ray diffraction (XRD) method on a single-exposure position sensitive scintillation detector (PSSD). Our experimental results were compared with (1) predictions made by assuming Tresca's yield criterion, and (2) a finite element model which treats the swage autofrettage process as a ram, mandrel, and cylinder three-body problem.

SPECIMEN PREPARATION

Figure 1 is a schematic diagram of the swaging process including the carbide tool. The cylinder under investigation had an outside radius of 15.69 cm (6.18 in.) and an inside bore radius of 5.69 cm (2.24 in.), giving an OD/ID ratio of 2.75. Radially forged cylinders underwent a number of finishing operations during the

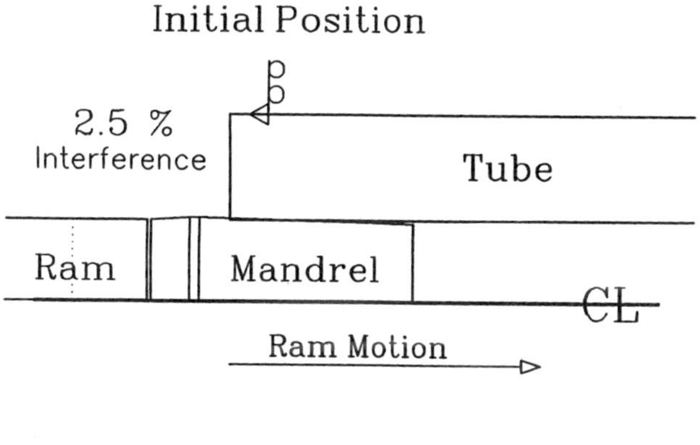

Fig. 1 - Partial view of ram-mandrel-cylinder swage autofrettage geometry.

manufacturing process. The inner bore of the forged cylinder was honed. Subsequently, the entire cylinder was subjected to plastic deformation by pushing a tungsten carbide tool through the bore. The interference between the ID of the cylinder and the load diameter of the carbide tool, which is 2.5 percent, determines the amount of plastic deformation. The inside radius of the cylinder was 5.69 cm (2.24 in.) before autofrettage and 5.77 cm (2.27 in.) after autofrettage. The bore was rough-machined and a substantial portion of the tube was removed to eliminate 'end effects' during the swage process. The wall variation due to the boring process was found at this stage, and the cylinder was rejected. The after-processing wall thickness was 9.93 cm (3.91 in.) at 0 degree (thickest) and 9.68 cm (3.81 in.) at 180 degrees (thinnest). This gave a maximum wall variation of 2.54 mm (0.1 in.) and an eccentricity of 1.27 mm (0.05 in.).

A 3.81-cm (1.50-in.) thick ring was cut from the cylinder and was machine- and hand-polished. Electropolishing of the entire cross section of the ring was done by the Chrome Plating Facility of Watervliet Arsenal (Watervliet, NY). The polishing solution was a mixture of 50 percent sulfuric and 50 percent phosphoric acid heated to 130°F, the anode was a tank lead piece, the voltage was 7 to 8 V, the current was 2 amp/in.^{2}, and the cathode-to-anode distance was 10 to 15 cm (4 to 6 in.). No external agitation device was used. It took 45 minutes to remove 5 mils from the surface of the ring. Surface material was removed so that sanding, machining, and oxidation effects would not influence the residual stress measurements.

EXPERIMENTAL METHOD

A D-1000-A Denver X-Ray Instruments Model stress analyzer was based on a Ruud-Barrett PSSD system. The instrumentation and calibration procedures for the analyzer are described in References 4 and 5. The instrument calibration curve using a four-point bend is shown in Figure 2. The system utilizes a chromium target tube,

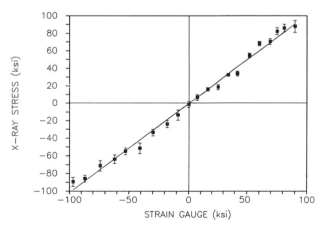

Fig. 2 - Four-point bend calibration of x-ray diffraction stress analyzer.

and the K-alpha reflects from the 211 plane of the body-centered-cubic steel at 2θ equal to 156.41 degrees.

Although local multiple-exposure (sin²ψ) software has been developed, the present measurements exclusively used the single-exposure method. This was due to the large number of measurements required to cover the 1-foot diameter surface. The surface was sectioned into 32 equal parts at an 11.25-degree span each. Data acquisition was set at three iterations at 2 seconds each. The data presented were the average of three measurements at each point on the surface of the specimen. For stress distribution determination, a specimen holder with a micrometer slide was used. Angular stress variation measurements were made by manually rotating and positioning the specimen. Taking into account the dispersion in the data and the alignment error, focusing error, and effects due to surface irregularities, the stress values were expected to have an error range of ± 10 to 15 Ksi. An IBM AT computer was used for data acquisition, control, and analysis. Symphony software was used for further data analyses and graphics.

X-RAY RESIDUAL STRESS MEASUREMENTS AND PREDICTIONS BASED ON TRESCA'S YIELD CRITERION

Hoop and radial residual stresses at 50 and 100 percent overstrain conditions based on Tresca's yield criterion are shown in Figure 3. Tresca's theory assumes that the thick-walled cylinder is overstrained by direct internal pressure and that an open-end condition is present [6]. The results were obtained by setting up a Symphony spreadsheet program with varying ID, OD, and percentage overstrain parameters. Because of the different stress conditions to induce the overstrain in the swage method compared with the direct pressure method, the actual stress distribution may differ. In Figure 4, theoretical hoop residual stresses, including the Bauschinger and hardening effects, are shown for typical thick-walled cylinders with OD/ID ratios equal to 2 and 3 [7].

The measured hoop and axial stresses versus radial distance at 0, 90, 180, and 270 degrees are shown in parallel plots in Figure 5. The 0-degree measurement position is where the ring is the thickest at 6.02 cm (2.37 in.), and the 180-degree measurement position is where the ring is the thinnest at 5.77 cm (2.27 in.). The plots show that all four curves follow the same common features as the compressive stresses that are observed at the bore and tensile stresses that are observed at the OD. Furthermore, the Bauschinger effect is observed in all four plots, with the 180-degree plot showing the steepest reduction of compressive hoop residual stress near the bore and coupled reduction of tensile stress near the OD.

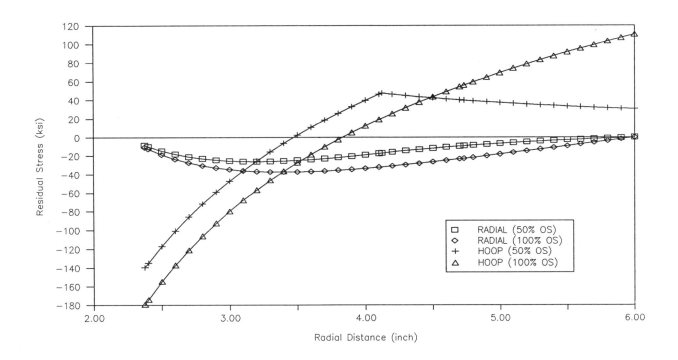

Fig. 3 - Theoretical hoop and radial stress distribution along the cylinder radius predicted by assuming Tresca's yield criterion.

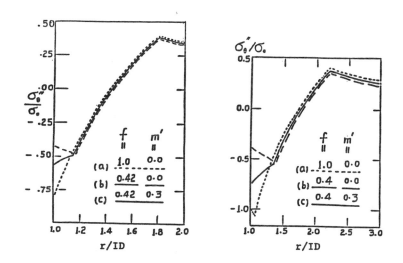

Fig. 4 - Residual stress distribution including Bauschinger (f) and hardening (m') effects in two autofrettaged cylinders [7] (left: OD/ID = 2, ρ/ID = 1.8; right: OD/ID = 3, ρ/ID = 2.2).

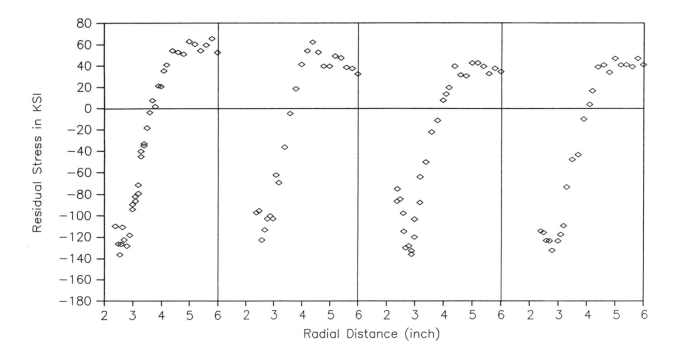

Fig. 5 - Hoop residual stresses along the radius at 0 degree (thickest section), 90 degrees, 180 degrees (thinnest section), and 270 degrees of the M256 eccentric cylinder.

Figures 6 and 7 show hoop and axial residual stresses at 0 degree and 180 degrees, respectively. In Figures 8 and 9, hoop stresses at 90 and 270 degrees, respectively, are given. The theoretical predictions assume 75 percent overstrain condition and are in fairly good agreement with our experimental results. The deviations in hoop stresses can be attributed to reverse yielding effects. The large ID hoop stress difference (-110 Ksi at 0 degree, -75 Ksi at 180 degrees) might justify the cylinder rejection, pending further experimental and theoretical verifications. Similar large deviations in hoop stresses in a cylinder have been reported in the literature [8]. Radial stresses do not quite converge to zero at the OD. It is interesting that the reverse yielding effects on hoop stress are observed in all four plots. At 180 degrees, which is the thinnest section of the cylinder, more pronounced effects exist.

Angular hoop stress variations at the ID and OD are shown in Figure 10. The measurements at the ID show a peak around 180 degrees and a smaller peak around 0 degree. The OD measurements do not show pronounced peaks. The angular stress results are in good agreement with measurements made on a Technology for Energy Corporation (TEC) stress analyzer [9].

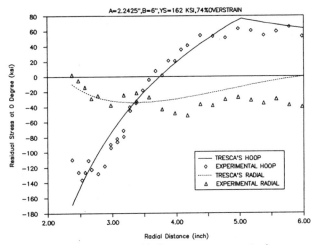

Fig. 6 - Hoop and radial stresses at 0 degree (thickest).

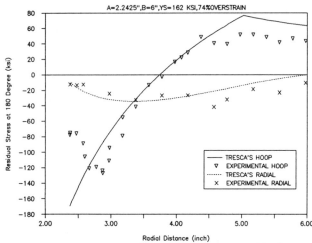

Fig. 7 - Hoop and radial stresses at 180 degrees (thinnest).

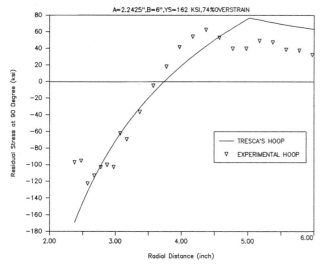

Fig. 8 - Hoop stress at 90 degrees.

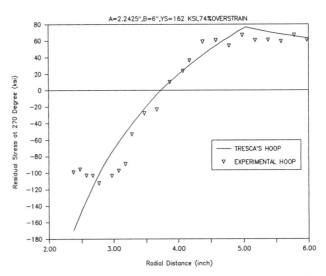

Fig. 9 - Hoop stress at 270 degrees.

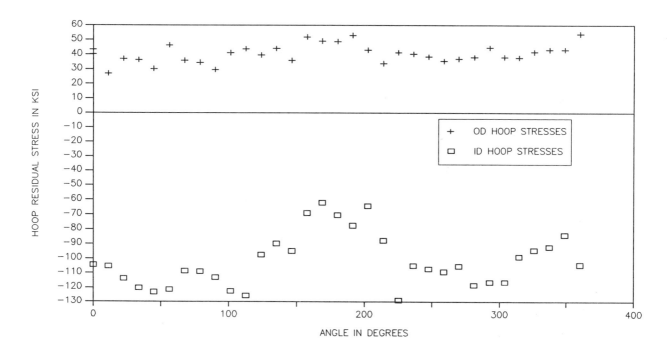

Fig. 10 - Angular hoop residual stresses around the cylinder.

ABAQUS FINITE ELEMENT MODEL

A finite element model using ABAQUS code was run on a Convex C-220. Residual stress analysis in cylinders can be accomplished by three different methods: (1) the use of classic closed-form equations assuming the plane-stress end conditions and elastic-perfectly plastic material properties; (2) the finite element analysis of a cross section using improved material definition, plastic deformation at the bore during unloading, and eccentric geometry; and (3) the full modeling of the complete problem with the mandrel sliding through the tube. This adds a great many more practical details including proper mandrel geometry, moving contact, and friction at the interface, and it also requires extensive computer resources. The second method was initially chosen to evaluate the magnitude of the eccentric bore effect.

The analysis was performed using the generalized plane-strain (CGPE10) elements to model one-half of the cross section of the tube at the time of the swage process. A total of

154 elements was defined using 7 rows in the radial direction and 22 rows in the angular direction. The eccentric geometers were produced by shifting the outer row of grid points relative to the inner row in increments of 0.10, 0.21, 0.50, and 1.0 cm (0.039, 0.078, 0.197 and 0.394 in., respectively). The load was applied by using fixed radial displacements at the inner row of nodes, which more closely approximates the action of the tungsten carbide swage mandrel than a uniform hydraulic pressure. The ABAQUS material stress-strain curve used was almost identical to results taken from an independent experiment for steel with about the same yield strength (162 Ksi or 1116 Mpa) as the production tube. The material was defined using five linear segments which fit the data in a least square sense.

As shown in Figures 11 and 12, the finite element model predicts little change in residual stresses due to reasonable values of eccentricity shown in decimeters. These are plots of hoop stress versus radius at the thickest (0-degree) and thinnest (180-degree) cylinder sections. It can be seen that the bore stresses are virtually the same. Farther at the outside of the tube, the eccentricity must be large to produce major effects. The hoop and radial stresses have the same features as the experimental results shown in Figures 5 through 9, except for the hoop stresses near the bore. However, the results clearly can not explain the experimental angular stress variation results shown in Figure 10. The predicted hoop and radial stresses for a symmetric cylinder using this finite element analysis are shown in Figure 13.

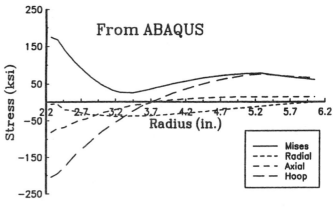

Fig. 13 - ABAQUS predictions of residual stresses of an M256 symmetric cylinder using the finite element method.

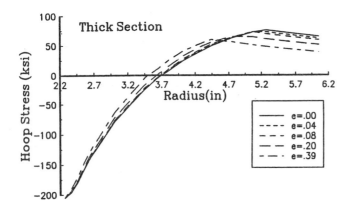

Fig. 11 - Finite element model hoop stress at 0 degree (thickest section) with varying eccentricity.

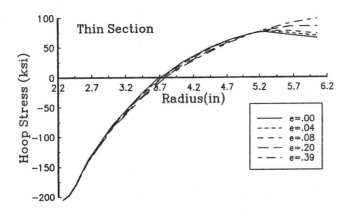

Fig. 12 - Finite element model hoop stress at 180 degrees (thinnest section) with varying eccentricity.

The differences between the measured and calculated hoop stresses near the bore have usually been attributed to the Bauschinger effect. An alternative finite element solution to the problem, using full finite element modeling of the whole cylinder as mentioned above, for a 105-mm cylinder is presented in Figure 14 [10]. This work was a full analysis of the swage process in a typical tube section and produced a residual hoop stress distribution much like the experimental data. This analysis was done without any reference to the Bauschinger effect and still showed a similar shape in the hoop stress distribution.

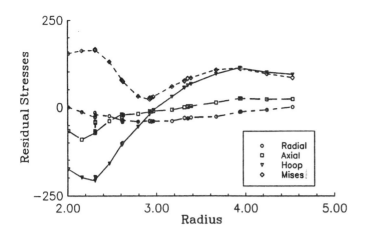

Fig. 14 - ABAQUS predictions of residual stresses of a 105-mm symmetric cylinder using an alternative finite element method.

SUMMARY

The following is a summary of our experimental and theoretical invesgitations of the eccentric swage autofrettaged thick-walled steel cylinder:

    1. Our XRD experimental results satisfactorily characterized the hoop and radial residual stress distributions of a swage autofrettaged cylinder.

    2. Our experimental results in residual stress distribution were in fairly good agreement with Tresca's theoretical predictions. The deviations could be accounted for by the Bauschinger effect. Further theoretical analysis in this area will continue.

    3. Our ABAQUS finite element swage autofrettage model predictions of residual stress distribution were in general agreement with our experimental values, except at the bore, where experimental hoop stresses are less compressive. Preliminary results using an alternative finite element analysis can predict the same residual stress behaviors near the bore as the Bauschinger effect.

    4. Our experimental angular distributions of hoop residual stress at the ID and OD showed peaks around 0 degree (thickest) and 180 degrees (thinnest). Our finite element model predicted very small variations in hoop and radial stresses even at large eccentricities. Variations in material properties, imperfection in the electropolishing process, and unknown effects due to the manufacturing processes may be the cause.

    5. Future investigations include experimental hardness determination and theoretical characterization of the cylinder using a refined finite element model.

REFERENCES

1. Davidson, T.E., C.S. Barton, A.N. Reiner, and D.P. Kendall, Exp. Mech. (February 1962).
2. Davidson, T.E. and D.P. Kendall, "The Design of Pressure Vessels for Very High Pressure Operation," WVT-6917, Watervliet Arsenal, Watervliet, NY, May 1969.
3. Rughupathi, P.S. and T. Altan, "Improvement of the Life of Carbide Tools Used in Autofrettage-Swaging of Gun Barrels," Final Report Contract No. DAAA22-79-C-0212, Battelle Columbus Laboratories, Columbus, OH, August 1980.
4. Ruud, C.O., Ind. Res. and Dev. 25, 84 (January 1983).
5. Lee. S.L., M. Doxbeck, and G. Capsimalis, "X-Ray Diffraction Study of Residual Stresses in Metal Matrix Composite-Jacketed Steel Cylinders Subjected to Internal Pressure," Proceedings of the 4th International Conference on the Characterization of Materials, Plenum Press, (in press).
6. Davidson, T.E., D.P. Kendall, and A.N. Reiner, Exp. Mech. 253-262 (November 1963).
7. Chen, P.C.T., J. Pressure Vessel Technol. 108, 108-112 (1986).
8. Clark, G., "Residual Stresses in Swage-Autofrettaged Thick-Walled Cylinders," Materials Research Laboratories Report MRL-R-847, Department of Defence Support, Commonwealth of Australia, 1982.
9. Lee, S.L., M. Doxbeck, and D. Snoha, "Comparison of Three Bi-Axial Stress Analyzers for Residual Stress Determination," U.S. Army ARDEC Technical Report, Benet Laboratories, Watervliet, NY, in preparation.
10. O'Hara, P., unpublished results, U.S. Army ARDEC, Benet Laboratories, Watervliet, NY.

Evaluation of Surface and Subsurface Stresses with Barkhausen Noise: A Numerical Approach

P. Francino and K. Tiitto
American Stress Technologies, Inc.
Pittsburgh, Pennsylvania

ABSTRACT

The evaluation of residual stress is of great interest since it can affect fatigue life. Many failures initiate at or near the surface. This criticality of surface integrity makes stress evaluations in these locations important.

The smallest of cracks cannot be tolerated in many steel components. Therefore, it is necessary to detect the precursor of cracks. As steel's strength increases, it becomes more brittle which leads to fracture critical conditions. In high strength steels, the critical crack size is small and the crack growth life is short. In fracture critical materials, the emphasis must be placed on the crack initiation process.

It is difficult to measure subsurface stresses in steels with conventional methods. X-ray diffraction requires that material be gradually removed and successive measurements made. This is due to the limited penetration of x-rays. In the magnetoelastic Barkhausen noise technique, penetration is considerably deeper. The depth of analysis for Barkhausen noise technique can be changed with instrumentation.

In this work, a mathematical model has been developed to predict the Barkhausen noise signal for various stress profiles. Two weighting functions were needed for the numerical work. The first weight accounts for the greater Barkhausen response that is produced by tension as compared to compression stress. The other weight accounts for the effect of the damping of the noise. This model demonstrates the suitability of the technique to detect subsurface stresses, whether tensile or compressive. To qualify the model experimental results were used to compare to calculated values.

DEPTH OF MEASUREMENT

In order to calculate the Barkhausen noise response for various stress profiles, the damping of the noise signal must be known. According to Tiitto (1), the damping of Barkhausen noise containing a distribution of frequencies between f1 and f2 can be described by a damping function D(x):

$$D(x) = \frac{\int_{f_1}^{f_2} g(f)\, e^{-A x_n f^{.5}}\, df}{\int_{f_1}^{f_2} g(f)\, df}$$

where $A = \sqrt{\pi \mu \sigma}$
 μ = permeability
 σ = electrical conductivity
 x = distance from surface

The depth of measurement is a result of the damping described by Equation (1) for the propagating electromagnetic Barkhausen noise. This depth can be determined by Equation (1) if material parameters are known. The damping is caused by eddy currents, experienced by the propagating electromagnetic fields created by magnetic domain wall movements. The effective depth of measurement is defined as the distance over which the signal is attenuated to 1/e of the original value (approximately 37%).

In this work, D(x) was computed as a function of depth by using a computer program for various values of permeability, conductivity and frequency range. The effective depth of penetration was determined from these calculations and used to further calculate the amount of Barkhausen noise detectable on the sample surface for various stress profiles. This depth of analysis was used as a limit of integration. The damping function was used as the nonlinear weight to account for signal attenuation.

According to Equation (1), the most important parameters affecting the depth of measurement are permeability, electrical conductivity and frequency range of Barkhausen noise selected for analysis.

The relative permeability is a ratio between the material permeability and the permeability of vacuum. It ranges between 100,000 for silicon-iron and 300 for 300M [2]. Electrical conductivity values for most steels fall between 10^6 and 10^7 /ohm*m, which were selected to represent the extreme cases [3]. Several frequency ranges of Barkhausen noise were used in this work.

Since Barkhausen noise cannot be regarded as random white noise, D(x) was calculated also for g(f) values varying from 10 down to 1 for the various frequency ranges. The results were close to those obtained when g(f) = 1, which was then used for this work. The actual values of depth of measurement may be somewhat higher than reported here due to the variations in g(f). The effective depth of penetration, x, obtained as described above can be plotted as a function of relative permeability ranging from 100 to 100,000 for two different conductivity values. Results are given in Figure 1. Based on results plotted in Figure 1, it is evident that the lower the permeability and electrical conductivity the higher the depth of measurement.

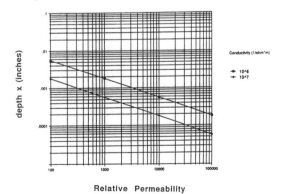

Figure 1. Effective depth of measurement as a function of permeability for two conductivity values of 10^6 and 10^7/ohm*m. Frequency range: 70-200 kHz.

APPLICATION TO M50 STEEL

M50 steel is a high temperature high speed tool steel used in aerospace bearing applications. D(x) calculated for M50 having a relative permeability of 104, electrical conductivity of $5*10^6$/ohm*m and several frequency ranges are given in Figure 2. This figure shows that the lower the frequency range of Barkhausen noise selected for analysis, the higher the depth of measurement.

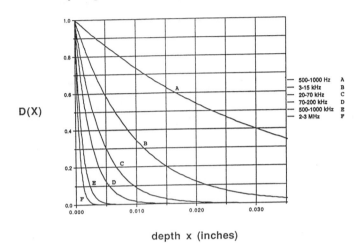

Figure 2 Damping of Barkhausen noise for M50 material and various analysis frequency ranges.

The effective depths of penetration, which corresponds to the value of 1/e for D(x) can be derived from these graphs and are tabulated in Table 1.

TABLE 1

Frequency Range	x @ D(x) = 1/e distance (inches)
500-1000 Hz	.03260
3-15 kHz	.00940
20-70 kHz	.00417
70-200 kHz	.00239
500-1000 kHz	.00101
2-3 MHz	.00055

Stress Profiles

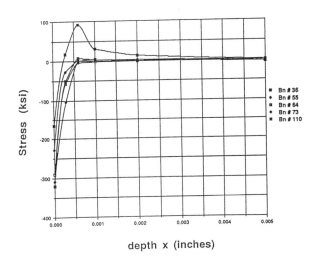

Figure 3   Typical Stress Profiles

X-RAY DIFFRACTION STRESS PROFILES

Typical stress profiles used in this work are shown in Figure 3 [7]. These profiles were obtained by x-ray diffraction on M50 components ground to different stress levels. To obtain the subsurface stress values, material was removed electrochemically and successive x-ray diffraction measurements were made. Corrections were applied to the measured stress values to account for material removal, stress gradients and x-ray beam penetration.

The depth of the layer from which a certain percentage of x-rays is diffracted can be calculated from Equation 2 [5]. TABLE 2 gives calculated depths (inches) of effective layers for M50 material studied. The linear absorption coefficient (μ) used in these calculations is 2490 (inch)^-1 [6].

$$x = \frac{\ln(1/(1-G_x))}{\mu\,(1/\sin(\Theta+\psi) + 1/\sin(\Theta+\psi))} \quad \text{EQ. (2)}$$

(hkl) = (211) with Cr K alpha radiation

where,

- x = Depth of layer corresponding G_x
- G_x = Percentage of x-rays diffracted
- μ = Linear absorption coefficient
- 2Θ = 154 degrees

TABLE 2
distance (inches)

Gx	$\psi = 0$	$\psi = 45$
95%	.00059	.00039
67%	.00022	.00015
50%	.00014	.00009

Table 2 indicates that the depth of measurement by x-ray diffraction is very shallow. According to Table 1, the Barkhausen noise method to determine residual stress is also limited to the sample surface due to the damping of the noise. Nevertheless, the depth of measurement with Barkhausen noise is a degree of order more than in x-ray diffraction.

WEIGHTED INTEGRATION

In order to calculate the amount of Barkhausen noise for various stress profiles, a weight to account for the variation in response to stress of the Barkhausen noise must be known. Typical stress calibration curves are shown in Figure 4. These curves are generated by strain gage response as compared with Barkhausen noise response in a simple uniaxial bending sample. The stress weight function was developed to have a similar form (shape) as a typical calibration curve. Several weight functions were tried to get better correlation between experimental Barkhausen noise response and calculated Barkhausen noise. Figure 5 shows the weight function used to account for variation in stress response.

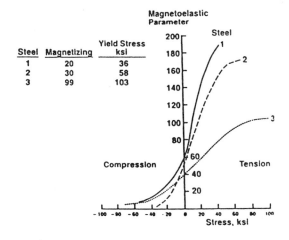

Figure 4.   Typical Stress Calibration Curves

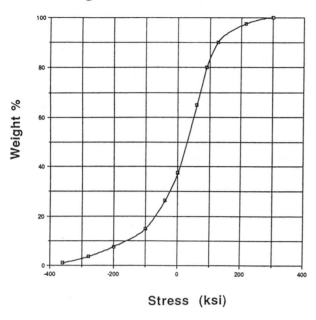

Figure 5. Weight Function for Stress

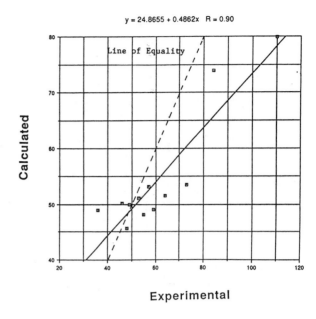

Figure 6. Correlation between Barkhausen noise calculated and experimental.

A computer program was written to convert stress profiles, such as those in Figure 3, into Barkhausen noise by using the developed weight functions and numerical integration. The program calculated the damping of these Barkhausen noise values by using the damping curve of Figure 2. This was done down to the calculated depth of penetration of 0.00239 inches for the M50 material under the experimental conditions applied. The Barkhausen noise values were calculated numerically using the integral shown in Equation (3).

$$BN_{cal} = C \int_0^{x\,@\,D(x)=1/e} W_{damping}(x)\, W_{stress}(s)\, dx \quad \text{EQ. (3)}$$

The calculated Barkhausen noise values for stress profiles are compared to measured Barkhausen noise in Figure 6. The experimental Barkhausen noise values were obtained on the samples before they were cut for x-ray diffraction stress measurements. The frequency range used was 70-200 kHz. Although our model underpredicted the experimental Barkhausen noise, a good linear relationship between experimental and calculated was found. The correlation between calculated and measured values is good, especially when accounting for the differences in the two experimental techniques.

PRACTICAL APPLICATIONS

Two practical applications will now be discussed. The first one is the stress generated in abusive grinding. Variation in grinding practice can produce drastically different stress profiles. Figure 7 shows three theoretical stress profiles that may results from grinding [8]. The numerical model developed in this work was used to calculate the corresponding Barkhausen noise levels. The calculated Barkhausen noise levels were 50 for "low stress" ground part, 100 for a "reground" part and 133 for a "severe" ground part. These results indicate that tensile surface or subsurface stresses can be detected by measuring on the surface of the component.

The second practical application is in shot peening evaluations. Figure 8 shows two theoretical stress profiles, one for not shot peened, the other for shot peened condition. The calculated Barkhausen noise levels were 50 and 30, respectively, which shows that compressive stresses can be determined with this technique.

It is important as we design and build with smaller margins and ask for greater than ever fatigue performance to control variation in residual stresses in today's components. The experimental and calculated results show that Barkhausen noise can help find variation in residual stresses which relate to component performance.

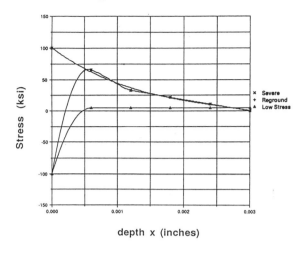

Figure 7. Theoretical Grinding Stress Profiles

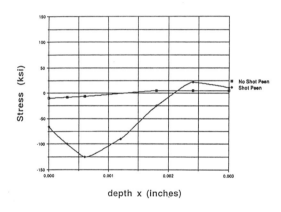

Figure 8. Theoretical Shot Peening Stress Profiles

REFERENCES

1. Tiitto, S. and Saynajakangas, S. Spectral Damping in Barkhausen noise. IEEE Transactions on Magnetics, 1975, MAG-11, 1666-1672.

2. Kraus, J.D. and Carver, K.R., Electromagnetics, McGraw Hill, NY, 1973, p. 201.

3. Metals Handbook, Vol. 1, American Society for Metals, Metals Park, OH, 1987, p. 150-151.

4. Cullity, B.D., Elements of X-Ray Diffraction, Addison-Wesley Publishing Company, Inc., Reading, MA, 1978

5. Hilley, M.E., ed., Residual Stress Measurements by X-ray Diffraction, SAE J784a SOCIETY of AUTOMOTIVE ENGINEERS, Warrendale, PA, 1971

6. Metals Handbook, Vol. 10, American Society for Metals, Metals Park, OH, 1986, p. 380-392.

7. Unpublished data from SPLIT BALL BEARING, Lebanon, NH

8. Gromly, M. W., Grinding Stresses; Cause, Effect and Control, Grinding Wheel Institute, Cleveland, OH

9. Prevey, Paul, THE SHOT PEENER Volume 4 Issue 3, Fall 1990

Application of a Unified Viscoplastic Model of Residual Stress to the Simulation of Autoclave Age Forming

S. Foroudastan and G. Boshier
Textron Aerostructures
Nashville, Tennessee

J. Peddieson
Tennessee Technological University
Cookeville, Tennessee

ABSTRACT

A unified viscoplastic constitutive equation has been combined with Bernoulli/Euler beam equations to create a model of autoclave age forming of beam specimens. Residual stress predictions based on the model are obtained numerically and compared with experimental data. Excellent agreement was found between simulations and observations.

Data obtained by several residual stress testing methods indicates that the autoclave age forming process, pioneered by Textron Aerostructures, will provide aluminum parts with lower residual stresses than more conventional forming methods.

TEXTRON AEROSTRUCTURES has pioneered the use of autoclave age forming to manufacture aircraft wing panels. In autoclave age forming a metal part is heated to a temperature sufficient for it to exhibit creep and stress relaxation. It is then subjected to pressure which forces it against a tool (mold). The part is held to the tool surface by the applied autoclave pressure for a period of time to allow stress relaxation to relieve the stresses produced by forming. Finally, the part is released and partially springs back to a shape somewhere between its original shape and the tool shape. The process is illustrated in Figure 1.

Autoclave age forming has many applications, especially in the aerospace industry. In order to be used effectively, however, the relation between final part shape and tool shape and final residual stresses must be known. Mathematical modeling represents one way to gain insight into this matter. This paper describes an attempt to simulate an experimental program of autocalve age forming of beam specimens carried out by Textron Aerostructures. The program is outlined below.

A series of experiments were performed in which straight rectangular beam specimens of 7075-T651 aluminum alloy were age formed into curved beams (See Figure 2). A range of forming pressure was employed. The specimens were first heated and then subjected to forming pressure. When contact with the curved tool was established, the forming pressure was maintained for 24 hours. The specimens were then cooled and the forming pressure was released. The springback and residual stresses were measured.

From the above statement it can be seen that, even for the geometrically simple case of age forming a straight beam into a curved beam, the process is quite complex. A variety of mechanical phenomena including elastic deformation, plastic deformation, creep, stress relaxation, and thermal stresses are potentially involved. The complexity of the mechanical responses which are involved makes the use of a unified model of elastic and inelastic behavior attractive.

In order to create a relatively simple model for this preliminary investigation, it was decided to neglect thermal stress effects, to assume that part/tool contact was achieved at all points simultaneously (thus eliminating the need to solve a contact problem), and to pick one of the available unified elastic/inelastic constitutive models.

This paper has three purposes. The first is to compare the residual stress experimental data which was obtained from autoclave age forming to other more conventional forming methods. The second is to report residual stress predictions in autoclave age forming and make comparison with the experimental data. The third is to demonstrate the application of a unified elastic/inelastic constitutive model to a practical problem.

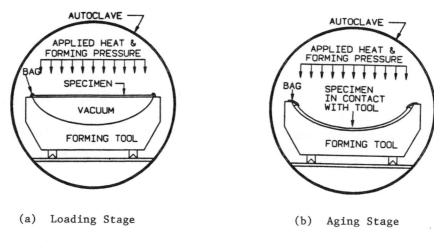
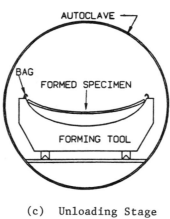

Figure 1 - Three stages of autoclave age forming.

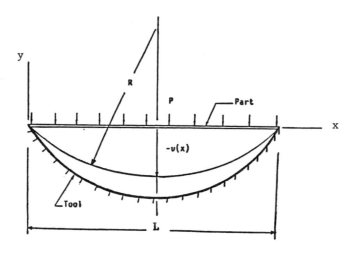

Figure 2 - Bar specimen and forming tool geometry.

EXPERIMENTAL RESIDUAL STRESS COMPARISON

One concern in forming contoured panels by cold mechanical forming methods is the amount of imparted residual stress left in the panel. If a high magnitude residual tensile stress is left in the material and the panel is subjected to cyclic loading, as it would be in aerospace applications, the possibility of inducing fatigue cracks is greatly increased. Therefore, the lesser the amount of residual stresses the better. Age forming offers a method of reducing imparted residual stress over other conventional methods. In an effort to substantiate this, 0.50 inch (1.27 cm) thick flat bar specimens were formed to a radius using several different forming methods and approaches. The methods included autoclave age forming, brake forming, brake forming with subsequent aging, roll forming, and roll forming with subsequent aging. The formed specimens were then measured for residual stress using the Hole Drilling Strain-Gage technique, as specified by ASTM E837-85. Figure 3 shows the results of these measurements. Clearly the age formed specimens contained less residual stress than specimens formed by the other methods. Therefore, imparting less residual stress to a contoured specimen during forming proves to be another benefit of the autoclave age forming concept.

CONSTITUTIVE MODEL

Several investigators have developed mechanical models capable of a unified description of yielding, creep, stress relaxation, nonlinearity and a variety of other mechanical effects. Some representative papers are those by Miller [1], [2], [3], Miller and Shih [4], Miller and Sherby [5], Hart [6], Bodner and Partom [7], Bodner and Merzer [8], Stoufter and Bodner [9], and Walker and Wilson [10]. A discussion of several currently popular models plus many additional references can be found in the paper by Janes et. al. [11]. The models discussed in the above mentioned references all have a similar structure involving evolution equations for measures of inelastic strain and a set of internal variables. Microscopic concepts can be used in determining the forms of the equations governing the evolution of the internal variables. The evolutionary nature of the methods allows a smooth transition from elastic to inelastic behavior and thus eliminates the need for the explicit inclusion of a yield criterion. It is felt that such models are potentially well suited for the simulation of a complicated process such as age forming in which the relative contributions of such phenomena as yielding, creep, and stress relaxation are initially unknown.

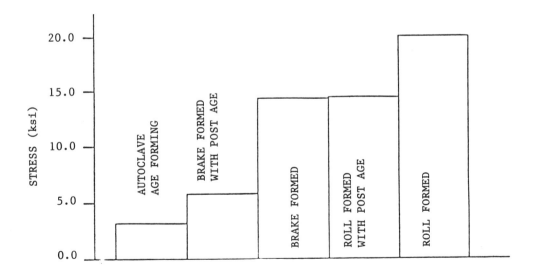

Figure 3 - Tensile residual stress measurements of aluminum alloy 7075 specimens formed to the same radius of curvature.

After investigating several possibilities the Miller/Sherby [5] model was finally selected. In a beam model only the one dimensional forms of the Miller/Sherby equations are needed. These are

$$\varepsilon = \sigma/E + e \qquad (1)$$

$$\dot{e} = B(\sinh(\sigma^2/(E^2 f)))^n \qquad (2)$$

$$\dot{f} = HB((\sinh(\sigma^2/(E^2 f)))^n - (\sinh(A^2 f))^n) \qquad (3)$$

$$f(0) = f_0 \qquad (4)$$

where a dot denotes time differentiation; σ denotes stress; ε denotes strain; e is called the inelastic strain; f is an internal variable (a dimensionless form of the quantity called the drag stress in [1-11]). The inclusion of the internal variable f provides a mechanism for roughly modeling the internal structural changes responsible for inelastic behavior. The quantities A, H, f_0, and n are material constants, while E and B are temperature dependent material properties. The physical interpretations of A, H, f_0, and n are discussed by Miller and Sherby [5]. The quantity E is the familiar Young's modulus of elasticity.

Substituting equation (2) into the derivative of equation (1) yields

$$\dot{\sigma} + BE(\sinh(\sigma^2/(E^2 f)))^n = E\dot{\varepsilon} \qquad (5)$$

In order to utilize the Miller/Sherby model to simulate the behavior of a specific material, a set of material constants appropriate for that material must be determined. In the present work this was done with reference to uniaxial constant strain rate tension tests, uniaxial constant stress creep tests, and uniaxial constant strain stress relaxation tests conducted by Textron Aerostructures on 7075-T651 aluminum alloy.

All the tests mentioned above involve uniaxial loading under quasistatic conditions. For these circumstances, the quantities σ, ε, and f can be regarded as constant throughout the specimen. The equations (3) and (5) are a set of two simultaneous differential equations which can be solved to determine either $\sigma(t)$ and $f(t)$ when $\varepsilon(t)$ is given (as in the tension and stress relaxation tests) or $\varepsilon(t)$ and $f(t)$ when $\sigma(t)$ is given (as in the creep tests). The problem of carrying out the solution of equations (3) and (5) numerically is complicated by the numerical stiffness of the system. In the present work it was decided to employ the Euler two point forward difference method and to deal with numerical stiffness by using a very small constant step size.

Because of the complicated structure of equations (3) and (5), it is impossible to identify each of the material coefficients with a unique aspect of material response. Thus, the determination of the coefficients involves a trial and error process based on a large number of simulations. (This is true for all unified models, as discussed recently by James et. al. [11]). For the sake of brevity, the details of the trial and error process employed herein are omitted. The values ultimately selected as appropriate for age forming simulations of 7075-T651 aluminum alloy were $A=3.75\times10^2$, $H=5.0\times10^{-3}$, $n=5.0$, $f_0=3.0\times10^{-9}$, $E=9.5\times10^3$ ksi (6.55×10^7 kpa), $B=1.65\times10^{-8}$ sec^{-1} (the latter two being appropriate for a temperature of 325 °F).

BEAM MODEL

To develop the equations necessary for beam-bending simulations let x,y plane be the plane of bending with the x axis being horizontal and the y axis vertically upward (See Figure 2). The neutral axis corresponds to y=0. For a simply-supported rectangular beam of length L, width b, and height h loaded by a downward uniform pressure P the maximum moment (occurring at the center of the beam) is

$$M = p\,bL^2/8 \qquad (6)$$

and the moment of inertia about the neutral axis is

$$I = bh^3/12 \qquad (7)$$

Since the stresses and strains vary from point to point on the cross section; the variables σ, ε, e, and f appearing in equations (1-5) must now be regarded as functions of both y and t and the dots must be thought of as partial derivatives with respect to time. Using the geometric relationship

$$\varepsilon = -y/R \qquad (8)$$

(where R is the radius of curvature), integrating equation (1) across the cross section, and using the definition of the bending moment

$$M = -2b \int_0^{h/2} \sigma y\, dy \qquad (9)$$

(where symmetry has been used) yields

$$\sigma = -3p\,L^2 y/(2h^3) + E\left((24y/h^3)\int_0^{h/2} ey\,dy - e\right) \qquad (10)$$

The simulation is carried out as follows. In the upper half of the cross section, N points were chosen and the quantities $\sigma_i(t)$, $\varepsilon_i(t)$, $e_i(t)$ and $f_i(t)$; i=1,2,....,N were associated with these points. Using the values of $\sigma_i(t)$ and $f_i(t)$ from the previous time step, equations (2) and (3) are used to compute new values of $e_i(t)$ and $f_i(t)$ at each point by the Euler method and corresponding $\varepsilon_i(t)$'s are found from equation (1). Then the new σ's are found from equation (10) (the integral therein being carried out by the trapezoidal rule).

AGE FORMING SIMULATIONS

The age forming process consists of three stages as discussed earlier. These will be called loading, stress relaxation and springback herein.

The loading stage was simulated exactly as described at the end of the previous section. It was terminated when the part was predicted to be in contact with the tool.

The stress-relaxation stage was simulated by holding R in equation (8) equal to the tool radius. Then, at each stage of the calculation, values of σ and f from the previous time step were used to solve equation (1) and (5) (with $\overset{\circ}{\varepsilon} = 0$) to compute new values of σ and f at each point on the cross section by the Euler Method. Then the corresponding values of e were found from equation (1) and the value of P from a combination of equations (6) and (9) (with the integral in the latter being performed by the trapezoidal rule). It should be recalled that during this stage the quantity P represents the difference between the applied pressure and the reaction between the specimen and the mold. This quantity decreases as stress relaxation occurs. The stress-relaxation stage was terminated after 24 hours.

The springback phase was simulated by rapidly reducing P to zero and holding it there. The order of calculation was the same as that previously described for the loading phase.

After some experimentation it was found that N=6 was adequate to produce meaningful results. Simulations carried out with this number of points took approximately 13 minutes of CPU time on a IBM 3090. The dimensions were assumed to be L=30 in. (76.2 cm), b=3 in. (7.62 cm), h=0.25 in. and 0.50 in. (0.635 cm, 1.27 cm). The initial state was assumed to be quiescent, that is

$$\sigma_i = \varepsilon_i = 0, \quad f_i = f_o; \quad i = 1,2,\ldots,N$$

The ability of the simulation procedure to estimate residual stresses is an extremely useful feature of the model, therefore this quantity was selected for graphical presentation. Residual stresses were predicted to exist in the final formed part.

Some typical residual stress prediction patterns are shown in Figures 4 and 5. These figures present normal stress versus elevation in the upper half of cross section of the specimen for different values of R_t/h (tool radius/thickness). It can be seen that as the thickness of the specimen increases, the maximum residual stress in the top layer of the convex side of the specimen increases. Also, it can be seen that the model predicts a consistent pattern of residual stress which is increasing as (R_t/h) is decreasing.

Experimental data was available only for $R_t/h = 100$ as shown in Figure 4. The experimental data was obtained by using the hole drilling strain-gage technique to a depth of 0.08 in. It can be seen that predictions agree well with the experimental data.

CONCLUSIONS

The foregoing described the development of a three part autoclave age forming idealized model for simple rectilinear beam specimens. The constitutive model was based on a mechanical approach in which yielding, creep, stress relaxation, and nonlinearity were the primary controlling factors. Initial verification of the constitutive model was performed by determining the material coefficients for aluminum alloy 7075-T651 (an alloy currently used for age forming application at Textron Aerostructures.

This three stage analysis of the autoclave age forming process was used to correlate residual stress after forming with previously determined residual stress values, obtained by physical measurement. Excellent correlation was demonstrated.

It was also shown that autoclave age formed specimens contained less residual stress than specimens formed by more conventional metal forming methods.

ACKNOWLEDGEMENT

The authors wish to thank the members of Engineering staff at Textron Aerostructures, for their encouragement and review of the manuscript.

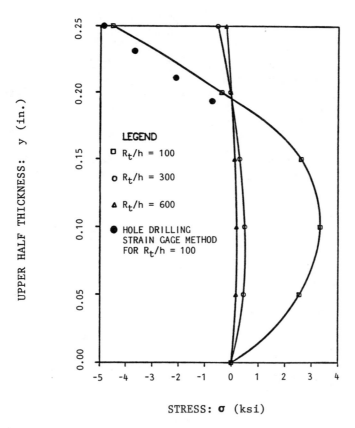

Figure 4 - Residual stress profiles for bending simulation of 7075 aluminum alloy.

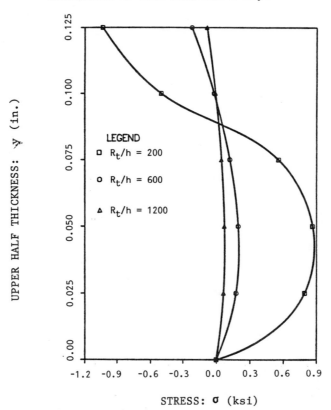

Figure 5 - Residual stress profiles for bending simulation of 7075 aluminum alloy.

REFERENCES

1 Miller, A.K., "An Inelastic Constitutive Model for Monotonic, Cyclic and Creep Deformation," Part I, ASME JOURNAL OF ENGINEERING MATERIALS AND TECHNOLOGY, Vol. 98, 1976, pp. 97-105.

2 Miller, A.K., "An Inelastic Constitutive Model for Monotonic, Cyclic and Creep Deformation." Part II, ASME JOURNAL OF ENGINEERING MATERIALS AND TECHNOLOGY, Vol. 98, 1976, pp. 106-113.

3 Miller, A.K., "Modelling of Cyclic Plasticity: Improvements in Simulating Normal and Anomalous Bauschinger Effects," ASME JOURNAL OF ENGINEERING MATERIALS AND TECHNOLOGY, Vol. 102, 1980, pp. 215-220.

4 Miller, A.K., and Shih, C.F., "An Improved Method for Numerical Integration of Constitutive Equations of the Work Hardening Recovery Type," ASME JOURNAL OF ENGINEERING MATERIALS AND TECHNOLOGY, Vol. 99, 1977, pp. 275-277.

5 Miller, A.K., and Sherby, O.D., "A Simplified Phenomenological Model for Non-Elastic Deformation: Prediction of Pure Aluminum Behavior and Incorporation of Solute Strengthening Effects." Acts Metallurgica, Vol. 26, 1978, pp.289-304.

6 Hart, E.W., "Constitutive Relations for the Nonelastic Deformation of Metals," ASME JOURNAL OF ENGINEERING MATERIALS AND TECHNOLOGY, Vol. 98, 1976, pp. 193-202.

7 Bodner, S.R., and Partom, Y., "Constitutive Equations for Elastic-Viscoplastic Strain-Hardening Materials," ASME JOURNAL OF APPLIED MECHANICS, Vol. 42, 1975, pp. 385-389.

8 Bodner, S.R., and Merzer, A., "Viscoplastic Constitutive Equations for Copper with Strain Rate History and Temperature Effects," ASME JOURNAL OF ENINGEERING MATERIALS AND TECHNOLOGY, Vol. 100, 1978, pp. 388-394.

9 Stouffer, D.C., and Bodner, S.R., "A Constitutive Model for the Deformation Induced Anisotropic Plastic Flow of Metals" International Journal of Engineering Science, Vol. 17, 1979, pp. 757-764.

10 Walker, K.P., and Wilson, D.A., "Constitutive Modeling for Engine Materials," United Technologies Pratt and Whitney Report FR-17911, 1983.

11 James, G.H., Imbrie, P.K., Hill, P.S., Allen, D.H., and Haisler, W.E., "An Experimental Comparison of Several Current Viscoplastic Constitutive Models at Elevated Temperature," ASME JOURNAL OF ENGINEERING MATERIALS AND TECHNOLOGY, Vol. 109, 1987, pp. 130-139.

Practical Applications of Residual Stress Technology, Conference Proceedings, Indianapolis, Indiana, USA, 15-17 May 1991

Modelling of Residual Stresses in Whisker-Reinforced Ceramic Matrix Composites

Z. Li
Argonne National Laboratory
Argonne, Illinois

R.C. Bradt
University of Nevada
Reno, Nevada

Abstract

Applying a modified Eshelby method, the residual stresses in SiC whisker/alumina matrix and sapphire whisker/mullite matrix composites have been calculated. The results are in excellent agreement with independent experimental measurements and confirm that significant residual stresses are present. Four factors are observed to influence the magnitudes and distributions of the residual stresses in these multiphase ceramic matrix composites. Those parameters are: (i) the thermal expansion mismatch, $\Delta\alpha$, (ii) the elastic modulus mismatch, ΔE, (iii) the geometry, (L/d) ratio, of the reinforcing phase and (iv) the concentration or volume fraction of the reinforcing phase.

ALMOST SINCE THE TIME that ceramics were first applied for utilitarian purposes, residual stresses in ceramics have been incorporated as a design parameter to improve their strength. Fine ceramic glazes with thermal expansions less than those of the underlying body are familiar to most. However cracked glazes frequently occur if the glaze does not develop compression during cooling. Ancient Chinese and Japanese potters took advantage of this cracking to develop the famous "crackle-glaze" patterns, which many collectors of fine porcelains now prize, centuries after their initial development. Surface compressive stresses to strengthen glass objects are now commonplace as most structural glass panels are tempered to enhance their strength and improve their safety. The latter, a consequence of the high strain energy density in the tempered glass, results because the fracture pattern of tempered glass is one of harmless, small equiaxed fragments as opposed to dangerous, large, lance-like pieces of dead-annealed glass. Materials scientists who are familiar with ceramics and glasses are thoroughly apprised of the above residual stress situations in commercial products.

Recently, materials scientists have developed ceramic matrix composites (CMC's) with ceramic whiskers as the reinforcing phases. These high-technology composites present a unique and interesting situation of very high levels of residual stresses. Because both constituents are "brittle" ceramics, there is little opportunity for annealing or stress relaxation to occur when these ceramic reinforced-ceramic matrix composites are cooled from their processing (synthesis) temperatures, which are often above 1000°C. This has the tendency to develop substantial residual stresses because of the large temperature differential (ΔT) on cooling from the processing temperature. Equally interesting is the feature that the reinforcing phases of these CMC's, be they whiskers or particles, are usually single crystals. This introduces the complications of anisotropic elasticity and anisotropic thermal expansion into the phenomenon of the development of the residual stresses. As single crystals often possess elastic moduli that may vary considerably with crystallographic orientation (A factor of two is not uncommon.) and thermal expansions can be equally anisotropic, it is evident that significant residual stresses may be expected to develop in these unique composites.

This paper addresses those residual stresses from a fundamental perspective, developing an Eshelby-type model for their theoretical calculation. After presenting the model, calculations are compared with independent experimental measurements. The model is subsequently applied to several ceramic/ceramic composite systems of current interest, including silicon carbide reinforced aluminum oxide and sapphire whisker reinforced mullite.

FORMULATION OF RESIDUAL STRESS CALCULATIONS

Studies of the residual stresses in multiphase composites have been in progress for several decades.

Numerous models have been proposed; however, among those various models, it is the modifications of the Eshelby method [1-3] which appear to be the most promising. This is because the Eshelby approach can exactly solve the stress field of an anisotropic second phase inclusion and is also relatively easy to apply to actual crystalline systems. Using the modified Eshelby method, the following residual stress problems in composites can be directly addressed and solved: (i) stresses within an anisotropic, single crystal, second phase inclusion, (ii) the stresses at the interface between a reinforcing second phase inclusion and the matrix, (iii) the average residual stress within the matrix, (iv) the effect of the geometry of the inclusion, (v) the effect of the volume fraction or concentration of inclusions and (vi) the change of the residual stresses within the reinforcing phase inclusions and in the matrix when external stresses are applied to the composite[4,5].

Figure 1 schematically illustrates an ellipsoidal inclusion within a matrix. The thermal expansions and the elastic properties of the inclusion are considered to be anisotropic, while those of the matrix are assumed to be isotropic. The descriptive geometric parameters of the ellipsoidal inclusion are denoted as L and d, where L is specified as parallel to the X_3 axis of the inclusion and d coincides with the X_1 and X_2 axes. The angle ψ in Figure 1 is the angle between the X_1 axis and the direction of interest. Two locations in the matrix just outside of the inclusion are of special interest. One is at the equator of the inclusion, $\psi=0°$, as denoted by point B and the other is at the pole of the inclusion, $\psi=90°$, as denoted by point A. If the (L/d) ratio of this ellipsoidal inclusion is equal to unity, then the ellipsoidal inclusion is a spherical grain. For an (L/d) ratio much less than one, the inclusion is a flat tabular plate, and when the (L/d) ratio is much greater than unity the inclusion approximates a whisker. Those three special geometric shapes of the inclusion are also illustrated in Figure 1 along the side of the schematic diagram.

Details of the mathematical formulation of the micromechanical stress calculations can be found in several references[2-7] and are presented only in the final matrix form here. To determine the internal stresses it is necessary to solve for the eigenstrain, ε^*_{ij}, which is generated by the thermal expansion and elastic moduli differences between the reinforcing second phase inclusion and the composite matrix. The eigenstrain of the inclusion is expressed as[4]:

$$[<C^D><S> + <C^M> - V_f<C^D><\bar{S}>]<\varepsilon^*> = <C^I><\varepsilon^T> - <C^D><\varepsilon^o>, \quad (1)$$

where the $<C^I>$, $<C^M>$ and $<C^D>$ are each 6 x 6 matrices of the elastic stiffnesses for the inclusion, the matrix, and the difference between the inclusion and the matrix,

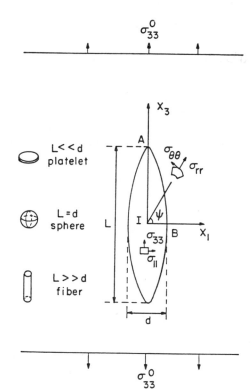

Fig. 1. Schematic of a Single Crystal Reinforcing Phase Inclusion in a Polycrystalline Ceramic Matrix.

respectively. The $<S>$ and $<\bar{S}>$ are each 6 x 6 matrices of the Eshelby tensor and the average Eshelby tensor, respectively. These are directly related to the geometry of the inclusion and the elastic properties of the matrix. The V_f is the volume fraction of the inclusion. The $<\varepsilon^o>$, $<\varepsilon^T>$ and $<\varepsilon^*>$ are each 6 x 1 matrices of the applied strain, the thermal strain and the eigenstrain, respectively. The ε^o_{ij} and ε^T_{ij} are given by:

$$\varepsilon^o_{ij} = (C^M_{ijkl})^{-1}\sigma^o_{kl}, \quad (2a)$$

and

$$\varepsilon^T_{ij} = (\alpha^I_{ij} - \alpha^M)\Delta T, \quad (2b)$$

where the $(C^M_{ijkl})^{-1}$ are the elastic compliances of the matrix. The σ^o_{kl} is the applied stress and the α^I_{ij} and α^M, respectively, are the thermal expansions of the inclusion and matrix. The ΔT is the temperature difference. Equation 1 can be considered to be the fundamental equation for the microstructural design of ceramic matrix composites on a mechanics basis.

The stresses within the reinforcing phase inclusion can then be calculated as:

$$<\sigma^{in}> = <\sigma^o> + <C^M>[<S> - <I> - V_f<\bar{S}>]<\varepsilon^*>, \quad (3)$$

where <I> is the 6 x 6 identity matrix. The average stress in the composite matrix is determined from:

$$<\sigma^M> = <\sigma^o> - V_f<C^M><\bar{S}><\epsilon^*>, \quad (4)$$

while the stresses just outside of the inclusion at the interface are expressed by:

$$<\sigma^{out}> = <\sigma^{in}> - <C^M><B><C^M><\epsilon^*> + <C^M><\epsilon^*>, \quad (5)$$

where <B> is a 6 x 6 matrix related to the elastic constants of the matrix and the unit vector outward from the inclusion[4].

From the above equations it is evident that the eigenstrain, ϵ^*_{ij}, is directly related to the internal micromechanical stresses. The eigenstrain and thus the residual stresses are influenced by the following four composite parameters: (i) the matrix - inclusion thermal expansion difference, $\Delta\alpha$, (ii) their elastic moduli difference, ΔE, (iii) the geometry of the reinforcing second phase inclusion, or its aspect ratio, (L/d), and (iv) the volume fraction of the reinforcing phase inclusion, V_f. The relationships between the eigenstrain and those four factors are clearly illustrated through Equation 1. The first and second terms on the right hand side of Equation 1, respectively, indicate the effects of $\Delta\alpha$ and ΔE on the residual stresses. On the left hand side of Equation 1 the <S> and V_f, respectively, indicate the effects of the geometry of the reinforcing inclusion, (L/d), and the concentration of the inclusions on the residual stresses.

Equations 1-5 indicate that although the residual stresses are highly dependent on the inclusion shape, those stresses are independent of the inclusion size. Many experimental results, however, show that the inclusion or individual particle size can greatly influence the mechanical properties of ceramic materials, such as strength, fracture toughness, etc.. Davidge and Green[8] and Kuszyk and Bradt[9] have clearly demonstrated the relationship between the microcracking and the inclusion or grain size of a particular composite material and non-cubic polycrystalline ceramic, respectively. They have proposed that when the strain energy of the inclusion or grain is larger than the surface energy required to form cracks, then microcracking may occur within the materials. Therefore, it is necessary to calculate the strain energy induced by the thermal and elastic mismatch within the composites. The formulation of the strain energy can also be readily derived from the modified Eshelby method[3]. In matrix form[4] it is expressed as:

$$W = W^o + W^* + W^T$$
$$= W^o + (1/2)V<\sigma^o>[<\epsilon^*> - <\epsilon^T>] - (1/2)V[<\sigma^{in}> - <\sigma^o>]<\epsilon^T>, \quad (6)$$

where W^o is the strain energy generated from the applied stress, W^* is due to the inhomogeneity of the inclusion and W^T is the thermal strain energy of the inclusion. V is the volume of the inclusion. Since the stress and the strain can be determined from Equations (1-5), the strain energy can be readily calculated. In the absence any externally applied stress, if the thermoelastic properties of the inclusion are isotropic and the shape of the inclusion is a sphere, then Equation 6 is identical to the formula derived by Davidge and Green[8]. It is evident from Equation 6 that the strain energy is not only dependent on the inclusion size, but it is also dependent on the shape of the reinforcing inclusion, (L/d), because both the stress and strain are shape dependent as well.

WHISKER REINFORCED CERAMIC MATRIX COMPOSITES

By applying Equations (1-5) the residual stresses within a SiC whisker reinforced polycrystalline alumina matrix composite and a sapphire whisker reinforced polycrystalline mullite matrix composite have been calculated. The thermoelastic properties of the reinforcing single crystal whiskers are treated as anisotropic while those of the polycrystalline ceramic matrix are considered to be isotropic. As illustrated in Figure 1 the reinforcing phase (whisker) will be assumed to have an ellipsoidal shape. The L dimension is defined to be parallel to the whisker growth direction. The growth direction of SiC whiskers is the [111]. For sapphire two different types, A-type and C-type, growth orientation single crystal whiskers are commercially available as reinforcing phases. The A-type sapphire whisker has an axial orientation of $<2\bar{1}\bar{1}0>$ while the C-type whisker has the <0001> growth direction. The thermoelastic properties of all of these materials are summarized in Table 1[10,11].

SIC WHISKER/ALUMINA MATRIX COMPOSITE - Since the SiC whisker reinforced polycrystalline alumina matrix composite has been demonstrated to exhibit significantly enhanced mechanical properties (strengths and toughnesses) relative to pure alumina[12,13], there have been several studies of the residual stresses in this material[4,14-17]. Figure 2 illustrates the residual stresses as calculated from the previous equations for a single SiC inclusion within a polycrystalline alumina matrix as a function of the inclusion (L/d) ratio. The stresses plotted in Figure 2 are the stresses within the single SiC whisker, σ^I_{ij}, and the radial and tangential stresses, σ_{rr} and $\sigma_{\theta\theta}$, just outside the inclusion in the matrix, as illustrated in Figure 1. The ΔT for the calculation has been chosen as 1000°C, for on cooling from the composite processing temperature the stress relaxation by creep processes will occur until around 1000°C. Figure 2 shows that the stresses within the SiC whisker σ^I_{ij} and the radial stresses σ_{rr} at the interface

Table 1. Thermoelastic Properties of Reinforcing Phases and Matrices

Single Crystal Reinforcing Phases								
	C_{11}	C_{33}	C_{12}	C_{13}	C_{44}	C_{14}	α_{11}	α_{33}
	(GPa)						($10^{-6}/°C$)	
β-SiC	352	---	140	---	233	0	4.45	---
Sapphire	497	499	164	112	147	-23.6	7.94	9.15
Polycrystalline Matrices								
	E		G		ν		α	
	(GPa)		(GPa)				($10x^{-6}/°C$)	
Alumina (Al_2O_3)	402		169		0.23		8.34	
Mullite ($3Al_2O_3 \cdot 2SiO_2$)	220		87.0		0.27		5.60	

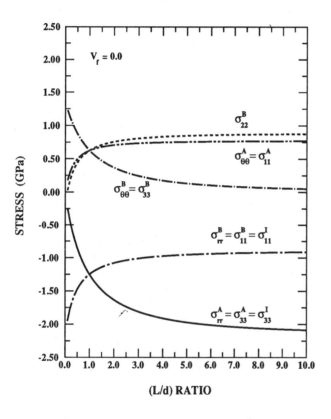

Fig. 2. Residual Stresses within a Single SiC Reinforcing Inclusion as a Function of the (L/d) ratio in an Alumina Matrix Composite.

are compressive (a minus sign) while the tangential stresses $\sigma_{\theta\theta}$ at the interface are tensile. The magnitudes of those stresses are significantly high, in the GPa range, and are highly dependent on the geometry of the inclusion, the (L/d) ratio. When the shape of the SiC inclusion varies from a flat plate to a whisker, as the (L/d) ratio is increased from 0.1 to 10, the σ_{33}^I continuously increases in compression from -0.25 to -2.0 GPa, while σ_{11}^I, which is equal to σ_{22}^I, decreases from -2 to -1 GPa. For all (L/d) ratios larger than about eight those stresses essentially remain the same. The tangential stresses at the interface are also dependent on the (L/d) ratio. As the (L/d) ratio increases, $\sigma_{\theta\theta}^A$ decrease to zero while $\sigma_{\theta\theta}^B$ and σ_{22}^B increase. Since the magnitudes of the residual stresses are very high, it is possible to create damage within the materials, as discussed in detail in reference [4]. It should also be noted that since the compressive radial stress or clamping stress, σ_{rr}^B, at the interface is extremely high, about 1 GPa, the SiC whisker and alumina matrix will be tightly bonded by these micromechanical residual stresses.

The effect of the volume fraction of the SiC reinforcing phase on the magnitude of the residual stresses can also be determined. Figure 3 depicts the residual stresses as a function of the volume fraction, V_f, of the whiskers for an L/d = 10. As the V_f increases the compressive stresses within the whisker are decreased while the tangential stresses at the interface and the average stress in the alumina matrix are both increased and remain in tension. The average stress in the alumina matrix increases from zero at $V_f = 0.0$ to 400 MPa at $V_f = 0.30$. This indicates that the volume fraction of the reinforcing phase is also an important factor for microdesigning any promising ceramic matrix composite, for the V_f directly affects the residual stress levels.

As indicated previously, the residual stresses within SiC whisker reinforced alumina matrix composites have been experimentally determined by different researchers. Predecki et al[14,15] have measured the residual stresses with different volume fractions of SiC by the X-ray diffraction technique. Majumdar et. al [16] and Tome et. al [17] have measured and analyzed the residual stresses by neutron diffraction techniques. Figure 4 compares the residual stresses experimentally determined by X-ray diffraction after Predecki et al. with the theoretical calculations previously outlined at different volume

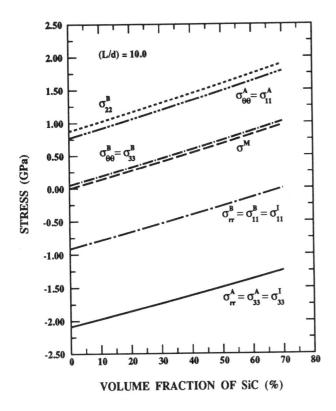

Fig. 3. Residual Stresses of SiC Whiskers, (L/d)=10, as function of Volume Fraction, V_f, for an Alumina Matrix Composite.

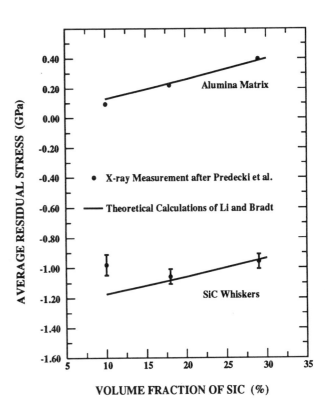

Fig. 4. Comparison of Average Residual Stresses between X-ray Measurements and Theoretical Calculations.

fractions of the reinforcing phase. Those residual stresses are the average stress within the SiC whisker, which is equal to $(2\sigma_{11}^I + \sigma_{33}^I)/3$, and the average stress within the alumina matrix, σ^M. Excellent agreement exists between the two independent studies, except for the stresses within the 10% V_f SiC whisker composite. This discrepancy is somewhat puzzling for it suggests that some stress relaxation processes may occur within these SiC whiskers, such as microcracking, or perhaps dislocation movement, because for the 10% SiC whisker composite the σ_{33}^I is about 2.0 GPa in compression, as illustrated in Figure 3. This stress relaxation may not significantly affect the magnitude of the average residual stress within the matrix, because the V_f of SiC whisker is small. However, since the magnitude of σ_{33}^I rapidly decrease as the V_f increases, which is also evident in Figure 3, stress relaxation may not occur for a V_f larger than 10% and the residual stress determined from the experiment and theoretical calculations are essentially the same. Figure 4 also reveals that no microcracking occurs inside the alumina matrix even for the $V_f = 0.29$, for which the average matrix stress σ^M is about 400 MPa.

The close agreement of the experimentally measured residual stresses and the theoretically calculated residual stresses is obvious in Figure 4. It confirms the use of the previously outlined Equations (1) through (6) to calculate the residual stresses for whisker reinforced ceramic matrix composites. Furthermore, it illustrates that when the second phase (reinforcing phase) of a composite is a single crystal (whisker or other shape), then it is necessary to apply concepts of anisotropic elasticity to calculate the residual stresses associated with the second phase. While these concepts have been specifically applied to the example of SiC whiskers reinforcing a polycrystalline alumina matrix, it is obvious that they are also applicable to many similar problems such as inclusions in steels and precipitates in age hardenable metal alloy systems. The aforementioned fundamentals of the residual stress calculations are applicable whenever a single crystal inclusion is present and the matrix can be treated as an isotropic material. Highly textured or large grain size matrices may not yield such close agreement.

SAPPHIRE WHISKER/MULLITE MATRIX COMPOSITE - Another ceramic whisker reinforced-ceramic matrix composite of technical interest for its potential long term high temperature thermodynamic stability is that of sapphire whiskers reinforcing a polycrystalline mullite matrix. Not only are the whiskers and matrix in chemical equilibrium to the melting temperature of mullite, > 1800°C, but the two phases are both oxides and thus stable in air. Unfortunately, the sapphire whiskers have a much

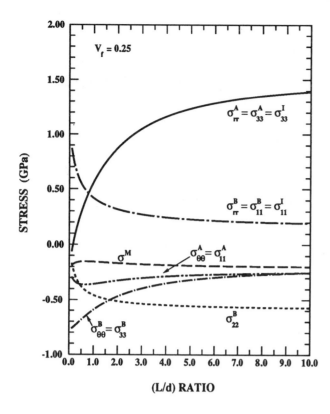

Fig. 5. Residual Stresses of 25% V_f C-type Sapphire Reinforcing Inclusion as a Function of the (L/d) Ratio in a Mullite Matrix Composite.

higher thermal expansion than the mullite matrix and this will result in the whiskers being in significant tension. The residual stress state will be the opposite of the previously discussed SiC whisker/alumina matrix system. There have not been any reported experimental measurements of the residual stresses in this interesting composite, but the close agreement of the experimental and theoretical results for the previous composite system suggests that the theoretical calculations are sufficient to define the residual stresses which will be present.

Figure 5 illustrates the residual stresses in a 25% V_f, C-type sapphire whisker inclusion within a mullite matrix composite as a function of the (L/d) ratio of the whisker. In contrast to the SiC/alumina composite the stresses within the sapphire whisker and radial stresses at the interface are tensile, while the tangential stresses at the interface and the average stress in the matrix are compressive. Those stresses are also highly dependent on the (L/d) ratio. When the sapphire inclusion becomes a whisker as the (L/d) = 10, the σ_{33}^I is about 1.4 GPa in tension while the average matrix stress, σ^M, is about 200 MPa in compression. The mechanical interfacial bonding stress, σ_{rr}^B, is lower, only about 200 MPa in tension. The residual stresses within an A-type sapphire whisker reinforced mullite matrix composite are similar to those of the C-type. However, since the thermal expansion along the $<2\bar{1}\bar{1}0>$ is slightly lower than that along the $<0001>$, the magnitudes of the residual stresses in A-type sapphire whiskers are lower than those in C-type whiskers. For a 25% A-type sapphire whisker composite the stress within the whiskers, σ_{33}^A, is about 900 MPa in tension and the average stress in the matrix, σ^M, is about 170 GPa in compression.

Mah et. al [18] have processed sapphire reinforced mullite matrix composites and measured their mechanical properties. The mechanically interfacial bonding stress in the sapphire/mullite composite is different from that in the SiC/alumina composite as the clamping stress around the sapphire whisker is tensile and the average stress in the mullite matrix is one of compression. If the fracture toughnesses are primarily influenced by the interfacial properties of these ceramic composites, then the fracture behavior of the sapphire/mullite composite should be superior to that of the SiC/alumina composite. However, the experimental results of Mah et al[18] reveal that the mechanical properties of the sapphire/mullite composites are not very good. The reason is illustrated in Figure 5. Large residual tensile stresses exist within the sapphire whiskers, which damage (fracture) the reinforcing phase and result in composite failure.

Li and Bradt[4,19] have discussed the residual stress effects on the fracture toughness of ceramic composites. In the SiC/alumina composite, although the interfacial bonding stress is high, which may not be favorable to the debonding and pull out, the large compressive stress inside the SiC whisker may inhibit crack propagation through the reinforcing phase and promote whisker bridging of crack faces in the following wake-region of a crack. Therefore, to design a promising composite material, the micromechanical stresses must be considered and understood.

SUMMARY AND CONCLUSIONS

A modification of the Eshelby method for the calculation of residual stresses in composites containing single crystal reinforcing phases has been presented. It addresses the anisotropic characteristics of the reinforcing phases, both the elastic anisotropy and the thermal expansion anisotropy. Stresses in both the reinforcing phases and the matrix can be calculated. The resulting methodology reveals that four factors significantly influence the magnitudes and distributions of the residual stresses in multiphase composites. These four factors are: (i) the thermal expansion mismatch, $\Delta\alpha$, (ii) the elastic modulus mismatch, ΔE, (iii) the geometry of the reinforcing phase and (iv) the concentration, or volume fraction of the reinforcing phase.

This model has been applied to several ceramic matrix composites, including SiC whiskers in a polycrystalline alumina matrix and sapphire whiskers in a mullite matrix. The results confirm that the residual stresses are indeed highly dependent on the composite microstructure, especially on the shape of the reinforcing phase microstructural constituents. For example, in the case of the SiC whiskers in the polycrystalline alumina matrix, the SiC whiskers are in a residual stress state of compression at room temperature. The longitudinal residual stress in the whiskers is highly dependent on the (L/d) ratio when the SiC whiskers are short, that is, for (L/d)<4. This is particularly significant for the residual compressive stress is very high in the 1 to 2 GPa range. For sapphire whiskers in a mullite matrix, very high tensile stresses develop in the whiskers and these may lead to the fracture of the sapphire whiskers during cooling from the high fabrication temperatures.

The aforementioned theoretical results have been compared with independent experimental measurements of the residual stresses. It is concluded that an Eshelby-type of analysis that fully accounts for the anisotropic elastic constants and thermal expansion coefficients is necessary to achieve accurate residual stress calculations when the reinforcing phase is present in single crystal form.

ACKNOWLEDGEMENT

The results presented in this paper were the product of research supported by NASA under Grant NAGW-199 and by US DOE, Basic Energy Sciences-Materials Sciences under Contract #W-31-109-ENG-38. The authors are grateful for the support of those agencies.

REFERENCES

[1]. J. D. Eshelby, *Proc. Roy. Soc. London, Ser. A*, **241** 376-396 (1967).

[2]. A. G. Khachaturyan, Theory of Structural Transformations in Solids, pp. 226-240, John Wiley & Sons, New York (1983).

[3]. T. Mura, Micromechanics of Defects in Solid, pp. 66-75, Martinus Nijhoff Publishers, The Hague, Netherlands, (1982).

[4]. Z. Li and R. C. Bradt, *J. Am. Ceram. Soc.*, **72**, 70-77 (1989).

[5]. Z. Li and R. C. Bradt, *J. Am. Ceram. Soc.*, **72** 459-466 (1989).

[6]. T. Mori and K. Tanaka, *Acta Metall.*, **21**, 571-574 (1973).

[7]. M. Taya, *J. Compos. Mater.*, **15**, 198-210 (1981).

[8]. R. W. Davidge and T. J. Green, *J. Mater. Sci.*, **3**, 629-634 (1968).

[9]. J. A. Kuszyk and R. C. Bradt, *J. Am. Ceram. Soc.*, **56** 420-423 (1973).

[10]. J. A. Salem, Z. Li and R. C. Bradt, pp. 37-43 in Proc. of Symp. on Advances in Composite Materials and Structures, Anaheim, CA, (1986), Edited by S. S. Wang, ASME, Fairfield, NJ. (1989).

[11]. K. S. Mazdiyasni and L. M. Brown, *J. Am. Ceram. Soc.*, **55**, 548-552 (1972).

[12]. P. F. Becker and G. C. Wei, *J. Am. Ceram. Soc.*, **67**, C267-C269 (1984).

[13]. M. G. Jenkins, A. S. Kobayashi, K. W. White and R. C. Bradt, *J. Am. Ceram. Soc.*, **70**, 393-395 (1987).

[14]. P. Predecki, A. Abuhasan and C. S. Barrett, Advances in X-ray Analysis, **31** 231-243 (1988).

[15]. A. Abuhasan, C. Balasingh and P. Predecki, *J. Am. Ceram. Soc.*, **73**, 2474-84 (1990).

[16]. S. Majumdar, D. Kupperman and J. Singh, *J. Am Ceram. Soc.*, **71**, 858-863 (1988).

[17]. C. N. Tome, M. A. Bertinetti and S. R. MacEwen, *J. Am. Ceram. Soc.*, **73** 3428-3432 (1990).

[18]. T. Mah, M. G. Mendiratta and L. A. Boothe, Technical Report **AFWAL-TR-88-4015** (1988).

[19]. Z. Li and R. C. Bradt, pp. 289-298 in Proc. Int. Conf. Whisker-and Fiber-Toughened Ceramics," Edited by R. A. Bradley, D. E. Clark, D. C. Larsen and J. O. Stiegler, ASM International, Oak Ridge, Tennessee, (1988).

Residual Stresses in Fuel Channel Rolled Joints in CANDU PHWRS

S. Venkatapathi
Atomic Energy of
Canada Ltd. Research
Chalk River, Ontario, Canada

T.A. Hunter
G.E. Canada Ltd.
Peterborough, Ontario, Canada

G.D. Moan
Atomic Energy of
Canada Ltd. CANDU
Mississauga, Ontario, Canada

ABSTRACT

A fuel channel in a CANDU[1] reactor consists of a zirconium alloy pressure tube, roll expanded into stainless steel fittings at both ends. Roll expansion of the joint results in residual stresses in the pressure tube and end fitting hubs. The effect of residual stresses on the integrity and reliability of the fuel channel has been extensively studied since the 1970s.

This paper discusses residual stresses in fuel channel rolled joints, techniques for measuring them and the effect of these stresses on integrity and reliability. In particular, the effect of rolled joint fabrication variables on the residual stresses are described, along with methods for optimizing these variables to minimize residual stresses. Also, two approaches for predicting residual stresses are covered based on dimensional measurements of the rolled joints, and development of an analytical model.

THE CALANDRIA VESSEL OF A CANDU REACTOR is a stainless steel cylindrical tank that contains the heavy water moderator. The calandria vessel is penetrated by 380 to 480 horizontal Zircaloy-2 calandria tubes connected to planar end shields. A fuel channel assembly consists of a zirconium alloy pressure tube, which passes through the calandria tube, and stainless steel end fittings.

The fuel channels contain the fuel and hot, high pressure heavy water coolant. Reactivity measurement and control devices, located in the low pressure and low temperature moderator, control the nuclear reaction. Figure 1 shows the salient features of a typical CANDU 6 reactor assembly.

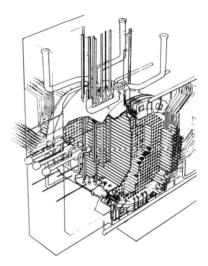

Fig. 1: CANDU pressurized heavy water reactor

The fuel channel, as shown in Figure 2, consists of a zirconium alloy (Zr-2.5% Nb) pressure tube, 103.4 mm (4.07") inside diameter x 4.1 mm (0.165") thick x 6.3 m (248") long, connected to stainless steel (AISI 403) end fittings, 165 mm (6.5") OD x 25 mm (1") thick, at each end by means of roll expanded joints. The pressure tube contains the fuel bundles and forms the portion of the fuel channel that resides within the reactor core.

---

[1] CANada Deuterium Uranium. Registered in the U.S. Patent and Trademark Office.

The end fittings form the out-of-core extensions of the pressure tube and provide connections to the fuelling machines (which perform on-power fuelling, and defuelling), and to the feeders connecting the fuel channels

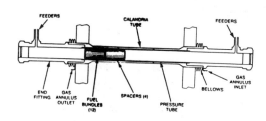

Fig. 2: Fuel channel assembly - schematic diagram

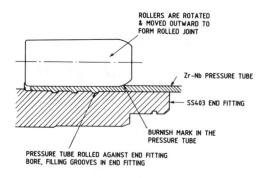

Fig. 3: Typical pressure tube to end fitting rolled joint (shown at the end of roll expansion)

to the remainder of the primary heat transport system. The end of each end fitting assembly is sealed by a channel closure, which is removed and replaced by remotely operated fuelling machines during on-power fuelling.

During the early stages of pressure tube reactor development, various methods of joining the pressure tube and end fittings were considered. The development of a sound and reliable metallurgically bonded joining technique between a zirconium alloy pressure tube and a stainless steel end fitting for operation in a near-reactor-core environment would require significant effort. This factor and considerations based on economy and the ease of replaceability of fuel channels favoured the use of roll expanded joints in CANDU fuel channels.

In forming a rolled joint (Figure 3), the pressure tube is rolled to a specified reduction in wall thickness, and the pressure tube material is forced into three circumferential grooves in the end fitting hub to provide a strong and leak-tight joint. The ability of the rolled joint to provide the required axial strength and leak tightness is confirmed by stress analysis and component testing.

The residual stresses in the pressure tube at the rolled joint region consist of (a) joint stresses, which occur in the roll expanded region of the pressure tube and are compressive, and (b) transition stresses, which occur between the rolled region of the pressure tube and the unaffected region of the pressure tube. The transition stresses result from the roll expanded region of the pressure tube pulling the un-rolled region of the pressure tube radially outward. The transition stresses can be either tensile or compressive depending on the dimensional variables in the roll expansion process.

High tensile residual hoop transition stress at the inside and outside surfaces of the pressure tube are not desirable because they can contribute to delayed hydride cracking (discussed later) in the pressure tube. The dimensional variables in the roll expansion process are controlled to result in a rolled joint with low residual tensile hoop transition stresses in the pressure tube.

The rolled-in compressive joint stresses in the pressure tube contribute to the interfacial stresses in the rolled joint, which in turn contribute to the axial strength and leak tightness of the rolled joint. However, the pressure tube protrusions in the end fitting grooves provide the more dominant contribution.

EFFECT OF RESIDUAL STRESSES

During the extensive investigations into the cause of the pressure tube cracks found in Pickering NGS A in 1974 and 1975 (References 1 and 2) and Bruce NGS A (Reference 3), more than 250 pressure tube to end fitting rolled joints were fabricated covering a wide range of fabrication variables. The objective was to identify the failure mechanism and its initiating conditions, and to determine methods of controlling and neutralizing the initiating conditions.

It was determined that the cracks in the pressure tube had grown by a process called Delayed Hydride Cracking (DHC). The DHC mechanism requires:
- a source of hydrogen and presence of hydrides,
- a driving force (stress gradient) to diffuse and concentrate the hydrogen and enhance precipitation of hydride,
- a large tensile stress to fracture the hydride.

The pressure tube material, (cold worked Zr-2.5% Nb is anisotropic and the crystallographic structure is such that its DHC behaviour is most sensitive to tensile hoop stresses in the pressure tube.

Control of the first two requirements to inhibit DHC occurrence in the pressure tube are discussed in detail elsewhere (Reference 5). This paper addresses the tensile stress aspects of DHC and the measurement of residual stresses in the rolled joint region. As well, the control of roll expansion process variables to minimize the tensile residual hoop stress in the pressure tube to prevent the occurrence of DHC is discussed.

The total stresses in the pressure tube are the sum of:
(i) operating stresses due to internal pressure, and thermal and mechanical loading, and
(ii) residual stresses in the pressure tube which are induced during fabrication of the pressure tube and roll expansion of the joint to the end fitting.

Operating stresses are limited by the jurisdictional and code requirements, as well as neutron economy considerations. Thus, the pressure tube fabrication, installation and roll expansion processes must be tailored to minimize the residual stresses.

Knowledge of the magnitude and distribution of the residual stresses in the rolled joint hub region of the modified AISI 403 stainless steel end fitting is required to perform a flaw tolerance evaluation of the end fitting.

The interface pressure between the pressure tube and the end fitting hub in the rolled joint region is of importance from leak tightness considerations. In three-grooved CANDU pressure tube rolled joints the interface stress is not a dominant contributor to axial strength of the rolled joint. When loaded axially to destruction, the rolled joint fails by the necking down of the pressure tube (Poisson effect) and the slipping and/or partial shearing of the pressure tube ridges out of the end fitting grooves.

PRESSURE TUBE RESIDUAL STRESSES

The zirconium alloy pressure tube is hot extruded from a billet and cold worked to about 25%. Then the outside surface is ground and the inside surface is honed to remove the surface layer. Finally, the pressure tube is stress relieved in a steam autoclave at 400°C for 24 hours which helps to relieve the residual stresses induced during fabrication. Only limited mechanical operations to achieve specified straightness and ovality are permitted after stress relief.

The rollers in the expander must be accurately positioned so that roll expansion of the pressure tube is well supported by the cylindrical bore of the end fitting hub. If the rollers are incorrectly positioned, such that the straight portion of the rollers extends into the tapered bore of the end fitting, high tensile residual hoop transition stresses are produced as shown in Figure 4.

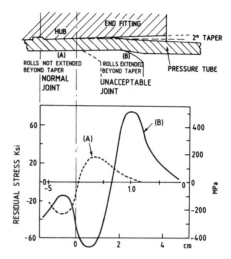

Fig. 4: Effect of roller position on residual hoop stresses at the inside surface of the pressure tube

The position of the roller is set so that the inboard ends (noses) of the rollers are well within the land between the inboard edge of the inboard groove and the start of the taper in the end fitting. During fuel channel installation in commercial power reactors, the "expander reach setting" is inspected and controlled closely. The burnish mark location in every rolled joint is inspected to ensure that it is within specified limits.

In rolled joints properly rolled as described previously, the pre-roll average diametral fit between the pressure tube and the end fitting affects the transition residual stresses in the pressure tube. A reduction in average diametral clearance decreases the maximum tensile transition residual hoop stress in the pressure tube as shown in Figure 5.

CANDU pressure tube to end fitting rolled joints are fabricated to a pre-roll average diametral fit of 0.05 mm (0.002 inch) clearance to 0.18 mm (0.007 inch) interference. The maximum tensile residual hoop stresses with this diametral fit range are between 55 MPa (8 ksi) compressive to 55 MPa (8 ksi) tensile and are acceptable from DHC considerations.

To assemble the pressure tube and end fitting with an interference fit, the end fitting is heated to a maximum temperature of 427°C using induction heating. The pressure tube is inserted into the end fitting and the components are cooled to ambient temperature before roll expansion.

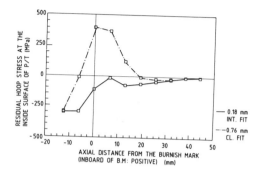

Fig. 5: Effect of average diametral fit on residual hoop stress at the inside surface of the pressure tube in a normal rolled joint

Seamless pressure tubes have a small variation in wall thickness. The circumferential variation may affect the circumferential distribution of transition residual stresses in the rolled joint. Consequently, the circumferential wall thickness variation in the region to be roll expanded is closely controlled.

Misalignment of the pressure tube and end fitting hub may also result in a larger residual stress variation in the pressure tube around the circumference, with the largest residual stress occurring at the largest clearance locations. During production, the pressure tube and the end fitting are carefully aligned before rolling.

RESIDUAL STRESS MEASUREMENT - PRESSURE TUBES

Analytical modelling of rolled joints using classical methods to determine residual stresses is inadequate to address the asymmetric nature of the stress field, and the anisotropy and work hardening effects of the pressure tube material. Such methods are cumbersome and cannot determine the magnitude and distribution of residual stresses in the pressure tube and end fitting with adequate accuracy. Thus, the determination of residual stresses and evaluation of the effect of rolling parameters on residual stresses have been addressed using experimental techniques.

In test programs, rolled joint assemblies are fabricated using equipment and procedures similar to those used for production fuel channel installation. The rolled joints in the fuel channel assemblies removed from the reactors during on-going surveillance or maintenance programs are also subjected to residual stress evaluation.

Residual stresses in the as-manufactured pressure tube, and the joint and transition stresses in the pressure tube after roll expansion are predominantly measured using a strain gauging and slitting technique.

Prior to strain gauge installation, the inside surface of the rolled joint is cleaned. Sets of "T" rosette strain gauges are installed in two axial strips 180° apart and two circumferential strips at the axial locations around the transition zone, just inboard of the burnish mark location. After baseline strain measurements are taken, the end fitting hub is slit and removed. The strains are then remeasured to determine the strain change due to hub removal. The outside surface of the pressure tube is now cleaned, and axial and circumferential strain gauge strips are installed on the outside surface at the same axial and circumferential locations as the inside strain gauges. The strains in this condition are read for both inside and outside strain gauges. The strains in the pressure tube regions with axial strips of strain gauges are relieved by cutting full length slots in the pressure tube. The strain changes are then measured and recorded.

The strains in circumferential strain gauges are relieved by making short axial slots and the change in strains are measured. Biaxial stresses are evaluated from these strains using elastic properties of the pressure tube. Figure 6 shows typical strain gauge installation and slitting patterns used to measure the transition residual hoop and axial stresses in the pressure tube after roll expansion. Figure 7 shows the axial distribution of residual stresses in a typical CANDU pressure tube rolled joint.

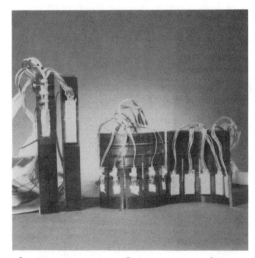

Fig. 6: Measurement of pressure tube residual stress at the transition zone by strain gauging and slitting

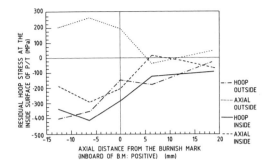

Fig. 7: Axial distribution of pressure tube residual stresses measured by the strain gauging and slitting technique in a CANDU pressure tube

Neutron diffraction offers a nondestructive alternative to stress measurements using strain gauging. The principle of stress measurement by neutron diffraction is identical to conventional X-ray diffraction stress measurement. Basically the measured quantity is lattice strain which is determined by precise measurement of lattice spacings using Bragg's law:

$\lambda = 2d_{hkl}\sin\theta_{hkl}$

where,
λ = Neutron wave length,
d_{hkl} = Lattice inter-planar spacing corresponding to a Bragg reflection (hkl) observed at a scattering angle $(\theta_{hkl})/2$,

and
(hkl) = Miller indices

The stresses are calculated from the measured strains using elastic constants.

The measurements are made using a neutron spectrometer operating in the diffractometer mode. A monochromatic beam of neutrons is obtained, from a beam of polychromatic neutrons extracted from the reactor, using a Ge crystal. A narrow neutron beam is then defined using slits in Cd masks. The beam is then passed through the sample and the scattering angle for diffraction is measured using a detector. The strain is determined from the measured change in the scattering angle. The experimental methods and analysis were developed in AECL Research (Reference 6) and in Europe (Reference 7).

The technique of neutron diffraction has been used to measure the residual stresses in the transition zone of a rolled joint rolled with incorrect roller position and has been demonstrated to be an acceptable technique. Not only is it nondestructive, it can measure the magnitude and distribution of radial stresses are measured in addition to the longitudinal and hoop stresses (Reference 8).

Photoelastic techniques and hole drilling methods have been used to measure the residual stress in the transition area of the pressure tube. The hole drilling technique has also been used to measure the residual stresses and the radial distribution of residual stresses in as-manufactured pressure tubes.

Maximum tensile residual hoop stress at the inside surface of the pressure tube can be predicted from profile measurements. In a pressure tube to end fitting rolled joint, longitudinal profiles of the inside surface of the pressure tubes are traced and measured at the burnish mark and transition zone as shown in Figure 8. Maximum tensile residual hoop stress is calculated using the following empirical correlation, derived by J. Van Winssen:

$\sigma_{IHmax} = AH + B\alpha$

where,
σ_{IHmax} = Maximum tensile residual hoop stress at the inside surface of the pressure tube
H = Flare Height (Figure 8)
α = Flare Angle (Figure 8)
and A and B are empirical constants.

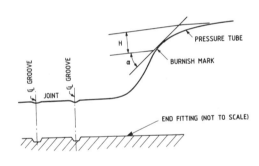

Fig. 8: Prediction of maximum tensile residual hoop stress at the transition zone at the inside surface of the pressure tube using the longitudinal profile of the inside surface of the pressure tube at the transition zone

Figure 9 compares the maximum tensile residual hoop stress predicted from profile measurements before strain gauging with those measured by strain gauging. The residual stresses derived by profile measurements represent the maximum tensile residual hoop stress just after rolling. During the operating life of a rolled joint in a reactor, the residual stresses will relax, but because of the constraints of the end fitting hub, the relaxation can be expected to occur at constant geometrical shape of the profile resulting only in an exchange of elastic strain for plastic strain. Profiling has been used to estimate the as-rolled maximum tensile residual hoop stress in the transition zone of a rolled joint during the post service examination of a rolled joint with a pressure tube with DHC occurrence (Reference 3).

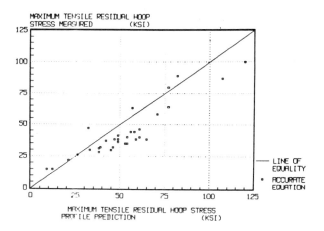

Fig. 9: Comparison of maximum tensile residual hoop stress at the transition zone at the inside surface of the pressure tube predicted using profiling technique with that measured using strain gauging and slitting technique

RESIDUAL STRESS MEASUREMENT - END FITTINGS

Knowledge of the magnitude and distribution of residual stresses in the end fitting is necessary for flaw tolerance evaluations and the calculation of interface stresses in a pressure tube to end fitting rolled joint. As explained in a previous section, these are determined by experimental measurements on rolled joints fabricated using production procedures and equipment.

Residual stresses in the end fitting hub have been measured using the Sachs boring and turning method. Strain gauges are installed on the outside surface of the hub of the rolled joint. After baseline measurements of strains, the pressure tube is machined out of the hub and the strains are remeasured. The end fitting bore is machined in steps, with strain measurements made after each machining step. Similar strain gauges are installed on the outside surface of a second rolled joint (fabricated to the same roll expansion procedure as the first one) and the strain gauges are read. The pressure tube is then machined out of the end fitting hub and strains are measured. Strain gauges are then installed on the inside surface of the end fitting hub and strain measurements are taken. The outside surface of the end fitting hub is machined in steps with strain measurements taken after each step. The residual stress profile across the wall thickness of the end fitting is then calculated from both sets of strain measurements.

Neutron diffraction technique has been used successfully to calculate the magnitude and distribution of residual stresses in a development rolled joint assembly. The non-destructive nature of this technique and the ability to provide triaxial stress distribution renders it an attractive alternative to the Sachs turning and boring method.

FUTURE DIRECTIONS

Rapid advances in computational techniques and finite element modelling software during the latter half of the 1980s have allowed us to re-examine the use of analytical techniques to model the pressure tube to end fitting rolled joint. Incorporation of the roll expansion process, material anisotropy and an asymmetric stress field may now be possible. However, the development of a rolled joint model and its experimental validation are in their early stage of evolution.

CONCLUSION

The database on the residual stress distribution in CANDU pressure tube to end fitting rolled joints is based on extensive component testing of more than 250 rolled joints. The comprehensive residual stress evaluation has resulted in the specification and fabrication of extremely reliable rolled joints in the fuel channel assemblies installed in commercial power reactors, contributing to the high performance and availability of CANDU nuclear reactors.

ACKNOWLEDGEMENT

The work covered in this paper is a summary of efforts that span three decades by personnel at Ontario Hydro, General Electric Canada and Atomic Energy of Canada limited. Their assistance and contributions are greatly appreciated.

This paper is dedicated to J. Van Winssen (retired), General Electric Canada, whose extensive work on the investigation of stresses in CANDU pressure tube to end fitting rolled joints have been invaluable in understanding the behaviour of CANDU pressure tube to end fitting rolled joints, and I.M. Burnie (retired), AECL, whose extensive experience and insight on fabrication techniques have been instrumental in many improvements in CANDU pressure tube to end fitting rolled joints.

REFERENCES

1. Ross-Ross, P.A., Dunn, J.T., Mitchell, A.B., Towgood, G.R., Hunter, T.A., "Some Engineering Aspects of the Investigation into the Cracking of Pressure Tubes in the Pickering Reactors", Paper Presented at the Annual Congress of the Engineering Institute of Canada, Winnipeg, Manitoba, 1975 September 30.
2. Coleman, C.E. and Ambler, J.F.R., "Delayed Hydride Cracking in Zr-2.5%Nb Pressure Alloy", Reviews of Coatings and Corrosion, Volume 3, pages 10-107, AECL 6250, 1979.
3. Dunn, J.T., and Jackman, A.H., "Replacement of Cracked Pressure Tube in Bruce GS Unit 2", Paper Presented to the Canadian Nuclear Society in Toronto, Ontario, AECL 7537, 1982 June.
4. Price, E.G., Moan, G.D., Coleman, C.E., "Leak Before Break Experience in CANDU Reactors", AECL 9609, 1988 April.
5. Coleman, C.E., Cheadle, B.A., Ambler, J.F.R., Lichtenberger, P.C., Eadie, R.L., "Minimizing Hydride Cracking in Zirconium Alloys", Canadian Metallurgical Quarterly, Vol. 24, No. 3, pages 245-250, 1985.
6. Holden T.M., Root, J.H., Fidleris, V., Holt, R.A. and Roy, G., "Application of Neutron Diffraction to Engineering Problems", Material Science Forum, Vols. 27/28, pages 259-370, 1988.
7. Allen, A.J., Hutchings, M.T., Windsor, C.G., Andreani, C., "Neutron Diffraction Methods for the Study of Residual Stress Fields", Advances in Physics, Vol. 34, No. 4, pages 445-473, 1985.
8. MacEwen, S.R., Holden, T.M., Hosbons R.R., Cracknell, A.G., "Residual Strains in Rolled Joints", Proceedings of ASM's Conference on Residual Stress - in Design, Process and Material Selection, Cincinnati, Ohio, 1987 April 27-29.

Practical Applications of Residual Stress Technology, Conference Proceedings, Indianapolis, Indiana, USA, 15-17 May 1991

Detection of Subsurface Tensile Stress in an Aircraft Engine Mainshaft Bearing Using Barkhausen Noise

W.P. Ogden
Split Ballbearing
Lebanon, New Hampshire

ABSTRACT

Abusive grinding can create undesirable subsurface residual stress patterns which can degrade aircraft main engine bearing performance. M50 steel is a high speed tool steel with excellent high temperature hardness properties which is used almost exclusively for turbine engine ball and roller bearing applications. This material is susceptible to a form of overtempering during grinding which only produces subsurface tensile stress. Conventional nondestructive inspection techniques, such as surface temper etch and eddy current, will detect grind damage if accompanied by a microstructural or hardness change but are not sensitive to residual tensile stress. Barkhausen Noise inspection has been shown effective in detecting residual stress variations in other materials susceptible to grind damage. This paper reviews a practical application of Barkhausen Noise inspection used to detect subsurface residual tensile stress in an engine mainshaft M50 roller bearing.

ABUSIVE GRINDING can create undesirable residual stress patterns which can degrade aircraft mainshaft engine bearing performance. M50 is a high temperature, high speed tool steel which is the predominate turbine engine bearing material. The high temperature properties of M50 make the material susceptible to grind damage which can go undetected with traditional NDE (nondestructive evaluation) methods. Subsurface residual tensile stresses can be created during grinding which can result in premature bearing failure. Barkhausen Noise inspection is the only known method capable of reliably detecting subsurface tensile stress. Barkhausen Noise has proven useful as a process control tool and as an NDE method in the manufacture of critical M50 mainshaft engine bearings.

BACKGROUND

M50 steel is a high speed tool steel with excellent high temperature properties which is used almost exclusively for mainshaft turbine engine ball and roller bearings. M50 contains .80-.85% Carbon, 4.00-4.25% Chromium, 4.00-4.50% Molybdenum and .90-1.10% Vanadium and is normally Vacuum Induction Melted-Vacuum Arc Remelted

(VIM-VAR) processed to improve microcleanliness. Developed to withstand jet engine soakback temperatures up to 399°C (750°F), M50 steel is typically triple tempered around 538°C (1000°F). Material and process controls yield M50 bearing components with uniform hardness and microstructure.

Abusive grinding of hardened steel can create significant forms of surface damage such as (1) cracks, (2) rehardened zones and (3) overtempered zones. All result from overheating at the immediate point of grinding. Cracks can be readily detected by magnetic particle and eddy current inspections. Rehardening and overtempering occurs in localized areas where the heat generated in grinding has been high enough to alter the metallurgical structure or hardness either by forming untempered martensite, in the case of rehardening, or by overtempering the existing martensite. Both these defects in conventional steels can be detected by surface temper etch or eddy current inspections.

Surface temper etch inspection relies upon a microstructure change occurring to indicate grind damage. The etch inspection process consists of a weak nitric acid solution (1-3%) in water or alcohol into which finish ground components are dipped until the surface turns a uniform gray. Deviation in surface color from uniform gray, either light or dark, must be evaluated for potential damage.[1] Etching can be likened to a photographic developing process in that the inspector must properly expose the component to the acid to develop the contrast. Typically, the etched surface appearance may be correlated with metallographic inspection to assess the potential consequences.

Rehardening grind damage is caused by local heating probably above 816°C (1500°F), and results in retransformation of the material to untempered martensite. Metallurgical destructive inspection for rehardening generally reveals a white etching zone consistent with untempered martensite. The rehardened zone is a hard, brittle phase characterized by a slight increase in hardness and is often accompanied by cracking.

Overtempering grind damage is caused by locally heating above the tempering temperature but below the critical temperature to transform the material to austenite (then to untempered martensite). With conventional bearing steels such as 52100 steel, this type of grind damage can be observed as a dark etching region associated with a reduction in hardness. Overtempering of M50, either as a dark etching region or reduction in hardness, is never observed during surface temper etch inspection. The high tempering temperature apparently does not allow for the formation of a dark overtempered M50.

Eddy current inspection has been used extensively with aircraft engine bearings to detect both residual grind damage effects and grind damage missed during surface temper etch inspection. Eddy current inspection relies upon the principle of electro-magnetic induction. In this NDE technique, an alternating magnetic field is produced by an alternating current passing through a coil of wire. Eddy currents are small electric currents created by an alternating magnetic field. They travel in predictable closed paths through any metallic media. However, a discontinuity in the metal near the surface will disrupt the eddy current field.

While most eddy current inspection techniques utilize a major ring with a known defect, the inspection has detected rehardening in M50 and some residual grinding effects not visible with temper etch. Some test evidence suggests that variation in residual stress could be detected by eddy current inspection.

PROBLEM

The existing NDE methods cannot reliably detect all overtempering effects on M50. Neither hardness nor microstructure of M50 are affected by slight overtempering. However, slight overtempering caused during the grinding process can induce a tensile residual stress zone in the near surface area which will not be found with either the etch or eddy current method.

The presence of this type of tensile stress has been suggested by several authors as detrimentally affecting bearing fatigue life. Zaretsky suggested that compressive subsurface stress could improve life.[2] Kuhlman reviewed surface stress of M50 bearings in an effort to correlate residual stress with grinding condition and showed that tensile stress was related to abusively ground bearing surfaces.[3] Barton suggested that Barkhausen Noise signatures can correlate with M50 bearing service life.[4]

BARKHAUSEN NOISE

Barkhausen Noise is created in ferromagnetic materials by applying an alternating magnetic field and measuring the abrupt changes in the magnetization. These abrupt changes in magnetization are caused by domain wall movement. Any material characteristic which restricts the domain wall movement will affect Barkhausen Noise. Residual stress and microstructural changes in ferromagnetic materials can be nondestructively detected with Barkhausen Noise.[5]

Under controlled conditions, Barkhausen Noise can quantify the residual stress of a metal surface. Barkhausen Noise has been shown to increase with (1) increasing tensile stress and (2) decreasing hardness.[5] Burkhardt and Kwun related Barkhausen Noise to residual tensile stress created by grinding.[6] The increase observed in Barkhausen Noise could have been related to microstructural changes. Burkhardt and Kwun concluded that Barkhausen Noise could be used to detect grinding damage. Pro also demonstrated that Barkhausen Noise could detect grind damage in gear teeth and piston pins.[7] Fix and Tiitto described the application of Barkhausen Noise to detect grind damage on automotive camshafts and to improve grind process control.[8]

Barkhausen Noise can respond to essentially three measurable material characteristics: (1) hardness, (2) microstructure and (3) residual stress. With M50 bearing steel, both hardness and microstructure are tightly controlled. Bulk hardness must be HRC 61-64, but is typically HRC 62-63. Microstructure is controlled in terms of retained austenite, grain size, martensitic structure and carbide morphology. Therefore, variations in Barkhausen Noise response can be attributed to variations in residual stress. More precisely, variations in the Barkhausen Noise within the same manufacturing lot of bearing rings can be a relative measure of residual stress.

PRACTICAL APPLICATION TEST RESULTS

A manufacturing lot of jet engine mainshaft M50 roller bearing outer rings was found to exhibit high Barkhausen Noise measurements in the rolling contact zone. The Barkhausen Noise inspection was performed using a Rollscan 200-1 instrument from American Stress Technologies. The sensor was custom designed to cover the entire race width and was oriented to measure Barkhausen Noise in the circumferential direction. Barkhausen Noise is expressed as a numerical value, magnetoelastic parameter (MP).

Statistical analysis of the magnetoelastic parameter values show that the production lot of bearing rings exhibited a bimodal distribution (see Figure 1). The histogram shows two populations of MP values. The bulk of the rings remain in a relatively low zone while a small group exhibit MP values up to 3 times higher. Conventional NDE methods, surface temper etch and eddy current, were not able to detect a defect which would account for the relatively significant variation. Metallurgical analysis of samples from each group did not reveal hardness or microstructural variations which would explain the observed phenomena.

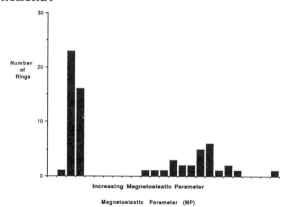

Figure 1 Barkhausen Noise Distribution- M50 Roller Bearing Outer Ring

Residual stress profiles performed at Lambda Research in Cincinnati, Ohio using conventional x-ray diffraction methods revealed significantly different subsurface residual stress distributions for rings exhibiting markedly different magnetoelastic parameters (see Figure 2). The residual stress measurements were conducted at depth intervals beginning at the surface, in order to detect tensile stress created by grinding, and were continued to a depth exceeding the estimated penetration of the Barkhausen Noise signal. Surface stresses of each sample were similar to each other and were consistent with nonabusive M50 grinding. However, the significant tensile stress, 569 MPa (82.5 KSI), at approximately .023 mm (.0009 in.) from the surface was abnormal. Normal subsurface stress distributions for M50 bearing raceways follow the general pattern of the ring exhibiting the lower MP value.

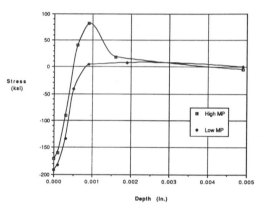

Figure 2 Residual Stress Depth Profile- M50 Roller Bearing Outer Ring

The residual stress depth measurements were taken in the circumferential direction, consistent with the Barkhausen Noise measurements. Material was removed by carefully electropolishing to the desired depth. The data presented has been corrected for both removal of material and penetration of the x-ray beam by Lambda Research calculations.

Bearing rings from the population exhibiting high Barkhausen Noise experienced an extremely high premature failure rate in turbine engine operation. Rig testing of selected rings from each grouping confirmed this effect on bearing performance. Residual stress profiles of selected rig tested rings verified the correlation of subsurface tensile stress to bearing performance.

A direct linear relationship between Barkhausen Noise, MP, and maximum subsurface tensile stress was not observed for these rings (see Figure 3). However, a threshold MP value was determined based upon limiting the maximum subsurface tensile stress to a safe operating level. Potential errors in the Barkhausen Noise measurement technique and residual stress analysis are under review in order to determine if the sub-surface residual stress pattern can be accurately predicted.

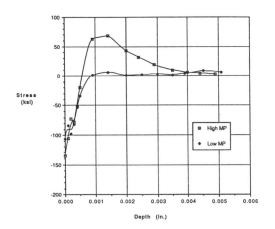

Figure 4 Residual Stress Depth Profile-M50 Roller Bearing Outer Ring

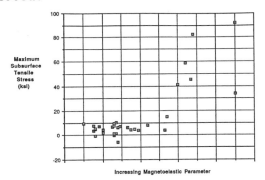

Figure 3 Barkhausen Noise Correlation With Subsurface Tensile Stress-M50 Roller Bearing Outer Ring

Subsequent production testing of M50 raceways has continued to validate Barkhausen Noise as a predictor of detrimental residual tensile stress. Eight different M50 bearing race configurations have been found to exhibit high MP values. In every case, subsurface tensile stress has been discovered to some degree (see Figures 4 and 5). In each case, the hardness and microstructure were investigated and found consistent with other similarly processed rings.

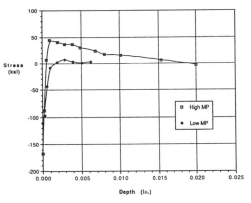

Figure 5 Residual Stress Depth Profile- M50 Roller Bearing Inner Ring

CONCLUSIONS

As a grinding process control tool, Barkhausen Noise has been valuable in assessing residual tensile stress. Grind process variables have been identified which affect the Barkhausen Noise response. The Barkhausen Noise results can be used to control the subsurface tensile stress in M50 bearing race grinding.

As an NDE method, Barkhausen Noise

has the potential of monitoring the residual tensile stress profiles in M50 bearing components. As currently used, Barkhausen Noise is an effective tool to detect a previous unidentifiable defect in M50 bearing steel, subsurface tensile stress. Improved bearing life resulting from consistent subsurface stress is to be expected from full exploration and application of Barkhausen Noise technology.

REFERENCES:

[1] Temper Etch Inspection, MIL-STD-867A, March 23, 1979, US Air Force, Washington, D.C.

[2] Zaretsky, E. V., "Selection of Rolling-Element Bearing Steels for Long-Life Applications," Effect of Steel Manufacturing Processes on the Quality of Bearing Steels, ASTM STP 987, J.J.C. Hoo, Ed., American Society for Testing and Materials, Philadelphia, 1988, pp. 5-43.

[3] Kuhlman, G. R., and Pardue, B. S., "Correlation of Residual Stress to Bearing Condition", Proceedings of ASM Conference on Residual Stress- in Design, Process and Material Selection, 27-29 April 1987, Cincinnati, Ohio, pp. 169-171.

[4] Barton, J. R. and Kusenberger, "Residual Stresses in Gas Turbine Engine Components from Barkhausen Noise Analysis," ASME Paper No. 74-6T-51.

[5] Tiitto, K., "Use of Barkhausen Effect in Testing for Residual Stresses and Material Defects," Proceedings of ASM's Conference on Residual Stress - in Design, Process and Material Selection, 27-29 April 1987, Cincinnati, Ohio, pp. 27-36.

[6] Burkhardt, G. L. and Kwan, Hegeon, "Residual Stress Measurements Using the Barkhausen Noise Method," Proceedings of the 15th Educational Seminar for Energy Industries, April 18-21, 1988, San Antonio, Texas.

[7] Pro, R. J., "Grinding Burn Detection During Production Using Magnetic Barkhausen Noise Measurements," Materials Evaluation, Vol. 45, June 1987.

[8] Fix, R. M., Tiitto S., "Automated Control of Camshaft Grinding Process by Barkhausen Noise," Material Evaluation, Vol. 48, July 1990.

Practical Measurements of Distortion and Residual Stresses Made by Welding and Pneumatic Hammer Peening Warm Weld Beads

R.W. Hinton
R.W. Hinton Associates
Center Valley, Pennsylvania

ABSTRACT

A simple device can be built by weld shops to measure weld distortions when comparing multibead weld deposits with and without pneumatic hammer peening of the warm weld beads. Residual stress measurements of as welded and warm peened weldments are presented for a 38mm (1 1/2-inch) thick ASTM A36 steel plate welded with E7018.

PNEUMATIC HAMMER PEENING of warm weld beads with a 6mm (1/4-inch) radius at the end of the hammer to relieve some of the weld distortion and residual stress has been successfully used for many decades by the welding industry to extend component and structural service life where cyclic mechanical or cyclic thermal stresses during service cause fatigue and cracking. A method developed by R. W. Hinton to allow a welder to measure the effect of his individual pneumatic hammer peening procedure and to produce the correct level of peening is described herein for E7018 structural steel welds. The test is designed to simulate weld and warm peening the weld beads between the root pass beads and cap beads.

To determine how much residual stress is relieved by using the proper weld and warm hammer peening procedure, residual stresses in two 38mm (1 1/2-inch) thick multibead welded ASTM A36 steel plates were measured.

Both 381mm (15-inch) long plates with 30° included angle double "V" joints and 1.5mm (1/16-inch) gaps at the root were shielded metal arc welded (stick) with 3mm (1/8-inch) E7018 electrodes. One plate sample was welded normally and the other plate sample was welded and pneumatic hammer peened while the weld bead was still warm.

A total of 26 weld passes were deposited in the down-hand, flat position. The plate section was rotated along the weld length from end to end to the other side after two weld beads were deposited on one side to balance the residual stress on both sides of the plate to prevent distortion. One plate sample was welded without peening. The other plate sample was welded and peened except that the two root passes and four cap beads were not peened to avoid cracking in the root passes and deformation and notching on the surface (cap beads). Ambient temperatures were about 24°C (75°F) and no preheat was used for the two samples. Interpass temperatures increased gradually and near completion of the welded sample the interpass temperature approached 260°C (500°F).

DEVELOPMENT OF A HAMMER PEEN GAUGE

DESCRIPTION OF PNEUMATIC HAMMER PEEN GAUGE PROCEDURE: A 9.5mm x 51mm x 305mm (3/8 x 2 x 12-inch) carbon-manganese steel sample, such as, ASTM A36 steel, AISI 1020 or other C-Mn steel is bolted on one end to a gauge plate as shown in Figure 1.

A longitudinal center-line is drawn on the test sample as shown in Figure 1 to deposit two 254mm (10-inch) long side by side weld beads with a 25% to 40% overlap.

The first steel sample is welded by depositing two 254mm (10-inch) long side by side weld beads that are 25% to 40% overlapped. Prior to depositing the next weld bead, the interpass temperature of the sample should not exceed 260°C (500°F). These side by side beads are four bead layers high.

After welding all eight beads and allowing the weld sample to cool below 260°C (500°F), the welder measures the deflection to the nearest 1.5mm (1/16-inch) at the free end between the top of the base plate and bottom of the 9.5mm (3/8-inch) sample.

Another 9.5mm (3/8-inch) thick steel sample is then bolted onto the gauge plate and welded in the following sequence:

1. Deposit two 254mm (10-inch) long side by side weld beads (#1 and #2) with a 25% to 40% overlap on the plate without peening and wait a few minutes to allow the sample to cool below 260°C (500°F) before depositing the next weld bead.
2. Deposit the third weld bead on top of the first (#1) weld bead and warm peen the third 254mm (10-inch) long bead between 538°C (1000°F) and 149°C (300°F). Wait a few minutes until the sample cools below 260°C (500°F) before depositing the next weld bead (Figure 2). (Use a hand-held pneumatic slag chipping hammer with a 6mm (1/4-inch) radius at the hammer end of the tool that has a 9.5mm to 12.7mm (3/8 to 1/2-inch) width.
3. Deposit the fourth weld bead on top of the second (#2) weld bead and warm peen as before between 538°C (1000°F) and 149°C (300°F). Wait a few minutes for the sample to cool below 260°C (500°F) before depositing the next weld bead.
4. Deposit the fifth weld bead on top of the third weld bead and warm peen. After the sample cools below 260°C (500°F), deposit the sixth weld bead on top of the fourth weld bead and warm peen.
5. Deposit the seventh weld bead on top of the fifth weld bead but **DO NOT PEEN** the seventh weld bead because this is a cap bead. Allow the sample to cool below 260°C (500°F) before depositing the eighth and last cap weld bead on the sixth weld bead without peening.
6. When the sample is below 260°C (500°F), measure the end deflection between the top of the gauge plate and bottom of sample to the nearest 1.5mm (1/16-inch) of the welded and peened sample after all eight weld beads are deposited. The desired welded and peened deflection should be less than the as-welded deflection by a factor of 1/3 to 1/2. Therefore welding and peening should reduce the as-welded deflection by 33% to 50%. For example, a 25mm (one inch) as-welded deflection is reduced by welding and peening to between a 17.5mm (11/16-inch) and 12.7mm (1/2-inch) deflection.

This procedure is applicable for structural steel manual welding rods from 2.4mm (3/32-inch) diameter and rod sizes in between and including 8mm (5/16-inch) diameter electrodes. Larger welding electrodes than 8mm (5/16-inch) diameter would require a thicker sample, such as a 12.7mm or a 16mm (1/2 or a 5/8-inch) thick 5mm x 254mm (2 X 10-inch) sample, but welding and peening is less affective for larger rods above 8mm (5/16-inch) diameter compared with smaller rods.

Comparison of the end sample deflection of a welded and peened sample with an as-welded sample deflection is shown in Figures 2 and 3, respectively.

If the welder does not achieve the 1/3 to 1/2 reduction in deflection of the original as-welded deflection, he can weld and peen another sample and adjust his peening procedure accordingly until a 1/3 to 1/2 reduction is achieved. It requires about one hour to weld each sample or a total of two hours to weld the first sample and to weld and peen the second sample.

DESCRIPTION OF HAMMER PEEN GAUGE: A sketch of the hammer peen gauge base is shown as reduced and not to scale in Figure 4. The 9.5mm (3/8-inch) thick hammer peened gauge sample is 51mm (2 inches) wide and 305mm (12 inches) long.

HAMMER PEENED RESIDUAL STRESS RELIEF OF WELD METAL

STRAIN MEASUREMENT: After double "V" welding a 38mm (1 1/2-inches) thick steel sample, the weld reinforcement was ground off, and 90° and 120° three leg rosette strain gauges were attached to the weld metal at a depth that was even with the plate surface of both the weld sample and welded and peened sample. The strain gauges were placed at a distance of 125mm (5-inches) in from the weld length end of each sample. The strain gauge section was then saw-cut from the sample until a 25mm x 25mm x 12.5mm (1 x 1 x 1/2-inch) thick section containing the attached gauge was band saw cut free of the plate sample. Principal residual stresses were then calculated from the strain relief measurements, and the results are listed in Table 1.

TENSILE PROPERTIES OF WELD METAL AND BASE STEEL: Results of duplicate tensile tests of the weld metal and base steel of the as-welded 38mm (1 1/2-inch) thick plate sample are listed in Table 2. Tensile properties of the weld metal corresponded to those expected of weld deposits using E7018 electrodes. Tensile properties of the base plate meet tensile properties of ASTM A36 steel plate.

Table 1. Weld Metal Residual Stress Measurements in As-welded And Peened 1 1/2-inch Thick Steel Samples

	Residual stress along the length of the weld	Residual stress across the length of the weld
Welded	413 MPa(60Ksi) tension	289 MPa(42Ksi) tension
Welded and Peened	294 MPa(43Ksi) tension	220 MPa(32Ksi) tension
Welded and Peened Reduction in Residual Stress	29%	24%

Table 2. Tensile Properties of 1 1/2-inch Thick Base Steel Plate and Weld Metal

	Yield Strength, MPa (0.2% Offset)	Tensile Strength, MPa	% Elongation	% Reduction in Area
Weld Metal	438(63.5 Ksi)	538(78.0 Ksi)	20.0	40.5
Base Steel	279(40.5 Ksi)	428(62.0 Ksi)	36.5	64.5

RESIDUAL STRESS RELIEF OF WELDING AND PEENING: The as-welded longitudinal tensile residual stress (413 MPa) approached the yield strength (438 MPa) of the weld metal as expected. In contrast, the welded and peened longitudinal residual stress (294 MPa) is about 67% of the yield strength of the weld metal.

In contrast to the weld and peen benefits in extending fatigue life, the influence of a 25% (Table 1 lists 29% to 24%) reduction in residual stress on the overload cracking resistance of furnace shells or vessels from hot-spot thermal-overload stresses is not enough to significantly improve resistance to cracking.

A reduction of longitudinal residual stress from 413 MPa tension to 294 MPa tension produces a better fatigue life of the weld metal. Fatigue stress amplitudes are limited to small values, such as, 20 MPa to 35 MPa when the residual stress approaches the yield strength. A fatigue stress amplitude of 70 MPa to 100 MPa is possible when the residual stress is reduced to 294 MPa tension, or about 67% of the yield strength of the weld metal.

In response to the amount of residual stress relief expected in the weld and peen repaired 57mm to 76mm (2 1/4-inch to 3-inch) thick steel shells, I offer the following comparison. There were 20 of 26 (77%) weld beads warm peened between the unpeened root weld beads and cap weld beads of the 38mm (1 1/2-inch) thick plate weld samples. Since about 90% of the weld beads of the 57mm to 76mm (2 1/4 to 3-inch) thick steel shells would be warm peened, the extrapolated percent of stress relief is above 30%. However, I recommend a 25% reduction in residual stress be used to make engineering estimates of improved fatigue life when properly welding and peening steel plate of all thicknesses ranging from 38mm (1 1/2-inch) and above. In general, the same weld and peen reduction in residual stress is expected for E60XX, E70XX, E80XX. Although not tested, higher strength weld beads above 550 MPa (80 Ksi) tensile strength may be less responsive to stress relief by welding and peening. This weld and peen technique is not commonly used for higher weld strengths using electrodes of E100XX and above.

ACKNOWLEDGMENT

The initial development of a hammer peen gauge procedure to train welders to properly weld and peen stainless steel was supported by R. W. Hinton's consulting work done for Revcon of Dedham, Maine. Expansion of this hammer peen gauge procedure to structural steel, and the measurement of residual stress relief by proper weld and peen techniques was supported by R. W. Hinton's consulting for Bethlehem Steel Corporation. Messrs. Mellet Wei and Barry Glassman of Bethlehem Steel Corporation provided the technical liaison with practical applications and provided review of this work. Messrs. R. Nelson and J. Nelson developed sample preparation and strain gauge measurement capabilities to successfully complete this study.

Figure 1. Photograph of Weld Hammer Peen Gauge (HPG) Base with 2 x 10 x 3/8-inch Sample Bolted to Gauge Plate.

Figure 3. Side View of Sample Deflection after Eight Weld Beads Were Deposited without Peening.

Figure 2. Photograph of Hammer Peened Third Through Sixth Weld Beads on HPG Sample.

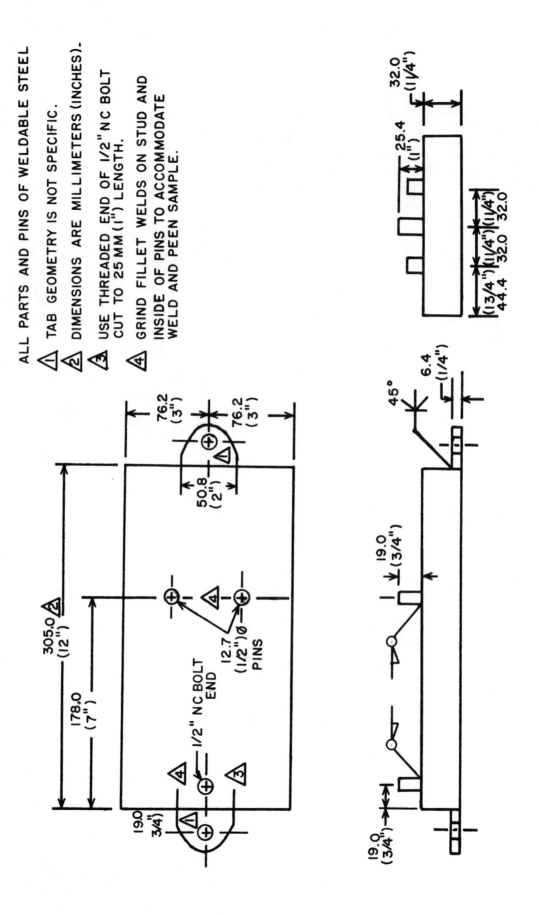

Figure 4. ENGINEERING DRAWING OF HAMMER PEEN GAUGE BASE

Intelligent Design Takes Advantage of Residual Stresses

J.S. Eckersley
Metal Improvement Company
Belleville, Michigan

T.J. Meister
Metal Improvement Company
Blue Ash, Ohio

ABSTRACT

Aerospace engineers, both in airframe and engines, discovered many years ago how to include residual compressive stresses in their designs to prevent or greatly retard catastrophic failures from metal fatigue, corrosion fatigue, stress corrosion cracking, fretting fatigue, etc. More recently, engineers in other industries are discovering how to use residual compressive stresses. In the automotive industries, beneficial stresses produce improvements in fatigue strength in the order of 30% on passenger car crankshafts; significantly reduce rotating mass in high performance engines; permit existing transmission designs to be upgraded for higher horsepowers without changing material or size of components.

Chemical industry engineers are using compressive stresses to overcome the detrimental effects of the HAZ of weldments in storage tanks and processing equipment subject to stress corrosion cracking and corrosion fatigue. Compressor engineers use beneficial residual stresses to increase, by 10 times, the life of very thin reed valves. Residual stresses are even used to generate the aerodynamic curvatures into wing skins, that are as much as 110 feet long, for the major commercial airliners.

The common denominator of all these examples of the use of residual compressive stresses is the process by which the beneficial stresses are generated. Controlled shot peening relies on the indentations of a metal surface, from the bombardment by millions of spherical particles, to introduce a high magnitude of residual compressive stress in that surface. Shot peening is an economical process that must be controlled and today computers are being employed to produce consistent results.

RESIDUAL STRESSES

It is a fact that mechanical engineering courses generally devote minimal attention to residual stresses in metals. This is unfortunate since, more often than not, residual stresses play a critical role in the survival or failure, depending on the sign of the stress, of a mechanical component. John Almen and Henry Fuchs, two of the pioneers in residual stress technology, encountered much opposition when they postulated that metals were capable of retaining what Fuchs described as "self stresses" (Ref. 1). Fracture mechanics finally validated their position and today those "self stresses" can be measured to a high degree of accuracy, if not convenience: the world is still waiting for the magic wand that will quantify residual stresses at depth, quickly, non-destructively, and in a shop environment.

In simple terms, metals experience only two types of stresses: applied and residual. Applied stresses are usually quite obvious: they are caused by the load applied to the mechanical member; from external forces such as weight or pressure or from assembly forces such as bolting or riveting. The rule is that if you remove the load, you remove the stress. The only subtleties in an applied stress system are the often overlooked stress concentrating effects of mechanical notches: sharp corners (inside and outside), holes, machine marks, rough surfaces, etc., (Fig. 1). An inappropriately placed notch can effectively take a generally low applied stress to beyond the yield strength or even the ultimate strength of the material, causing a premature failure.

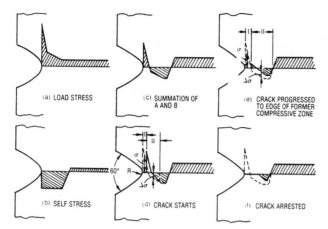

Fig. 1 - Crack Arrest By Compressive Stresses [A.].

On the other hand, residual stresses (or self stresses) remain in the metal, regardless of the load (unless the yield-point is surpassed or high heat has been applied). The source of residual stress is generally a common manufacturing process and in many cases the residual stress remains unnoticed and unaccounted for in stress calculations. However, these unseen stresses can have an enormous effect on the life of a mechanical component, depending on the sign of the stress (Table 1).

All stresses are either tensile (+) or compressive (-) in nature. Applied stresses are reasonably easy to visualize: if we apply a bending load to a metal bar, one side will be stretched in tension and the other side will be pushed together in compression. If the load is high enough, the bar will always break from the tension side, never from the side in compression. Residual stresses are much more difficult to visualize, particularly as to whether they are tension or compression, which is the reason Dr. Fuchs coined the term "self stress". The same cautions apply: residual stresses in tension contribute to premature failure and in compression contribute to greater part life. This is because residual tensile stresses **add** to the tensile stress of the applied load, while residual compressive stresses **subtract** from it (Fig. 2).

STRESSES AND MANUFACTURING PROCESSES

Unfortunately, many of the more usual manufacturing processes generate detrimental tensile stresses (see Table 1). Conventional grinding, for instance, can generate residual surface stresses that exceed yield or even the ultimate tensile strength of the material (Fig. 3). These exceedingly high stresses sometimes manifest themselves as what are known as "heat checks," micro cracks that are perpendicularly to the direction of grinding. Even if the heat checks are not visually present at

TYPICAL FATIGUE STRENGTH INCREASES

	Before Peening	After Peening
Coil Springs	75.0 KSI	125.0 KSI
Titanium Joint	38.0 KSI	84.0 KSI
Carburized Gears	31.0 KSI	73.0 KSI
HY80 Weldments	6.0 KSI	46.0 KSI
4340 Pins	71.0 KSI	87.0 KSI

100 KSI - 70 kg.mm²

TYPICAL FATIGUE LIFE INCREASES

	Cycles Stress Level	Cycles Before Peening	After Peening
Carburized Gears	80.0 KSI	200,000	30,000,000
Leaf Springs	140.0 KSI	300,000	NO FAILURE
4340 Polished	90.0 KSI	53,000	200,000
4340 Electroless Nickel Plated	90.0 KSI	39,000	141,000

100 KSI - 70 kg/mm²

Table 1

first, any applied tensile load will soon cause surface cracking to appear. If we think of a micro crack as a surface notch with an extremely high stress concentration factor (because of the very sharp "radius" at the tip of the crack), we will realize that just a few load cycles will bring on metal fatigue.

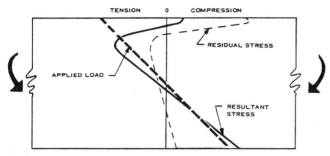

Fig. 2 - Resultant Distribution of Stress in a Shot Peened Beam with External Load Applied. Solid Line is the Resultant Stress. [B.]

Welding is another generator of very high residual tensile stresses. As the weld cools and shrinks, it pulls on the adjacent material in the heat affected zone, stressing it in tension of high magnitude. Weldments are often used in process equipment in the presence of corrosive elements. The combination of tensile stresses and even a slightly corrosive environment makes for a prime candidate for stress corrosion cracking (Fig. 4).

DESIGNING FOR COMPRESSIVE STRESSES

In both the cases of grinding and welding, as well as for many other situations of premature component failure, the solution is to change the sign of the stress in the surface of the component, before it is put into service. If the surface is yielded by cold working from rolling or shot peening, for instance, any surface residual stress will be replaced by residual compression. As long as the surface remains in residual compression, the part will not fail from fatigue, stress corrosion cracking or corrosion fatigue.

Controlled shot peening is the most convenient and economical process with which to generate beneficial compressive stresses since it is not geometry dependent. The process consists of bombarding the part with millions of cast steel spheres, (typically 0.023 inch /0.51 mm), each a tiny peening hammer. Sometimes, spheres of conditioned cut wire, glass or ceramic are used for special purposes. Impact from each bead causes a small impression or dimple in

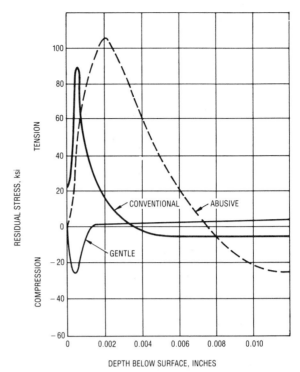

Fig. 3 - Residual Stress in 4340 Steel (HRC 50) After Surface Grinding. [C.]

the surface. Below each dimple is a hemisphere of cold worked material that has been yielded in tension and that sustains a corresponding residual compression as the cold worked material attempts to restore the surface to its original condition.

The resultant residual compression has a magnitude of at least 60% of the ultimate tensile strength of the material (Ref. 2). This 60% compression will offset the tension from an applied load, increasing the fatigue life, often by several orders of magnitude. In the case of stress corrosion cracking,

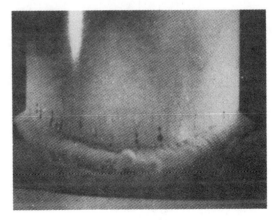

Fig. 4 - Stress Corrosion Cracking in Unpeened Type 321 Stainless Steel After 22 hr. in $MgCl_2$ Test. No Cracks Developed in Similar Peened Weldment After 264 hr. [B.]

the protection is virtually permanent unless there is aggressive general or pitting corrosion that will eventually eat through the compressive layer. Even in severe pitting, though, the shot peened component will last longer because the pitting will have to penetrate through the up to 0.040 inch/1 mm layer of compressed material before stress corrosion cracking can occur.

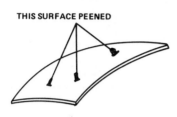

Fig. 5 - Compound Curvature Result of Tri-Axial Forces Induced by Compressive Stresses. [D.]

APPLICATIONS OF COMPRESSIVE STRESSES

Shot peening is used extensively in the aircraft industry on structural components, on landing gears, on engine components of every kind, in the high precision gears and shafts found in turboprop and helicopter transmissions, etc. Employing the principle of triaxial stresses, (Fig. 5) peening generates the aerodynamic curvatures in sculptured wing and empennage skins, doors, fuselage panels, stringers, etc., often to exacting contour tolerances. The same principles can be used to correct the shape of more mundane parts such as shafts and ring gears. The automotive industry, where shot peening was first used in the early 1930s, has finally begun to employ shot peening to its full potential. Now that cars are being designed with considerations for fuel efficiency and durability, shot peening is emerging as an elegant means of reducing the rotating mass in high performance engines, to permit much more compact transmissions, to balance comfort and performance in suspension systems.

Shot peening has found application in many other industries: a few examples will serve to illustrate the variety of components benefiting from the judicious use of residual stresses. The huge propellers for the world's largest cruise ships are shot peened against corrosion fatigue and cavitation. Thousands of Inconel tubes are peened "in situ" to prevent stress corrosion cracking on the 0.75 inch/20 mm internal diameter, in nuclear power station steam generators (Fig. 6). Aging

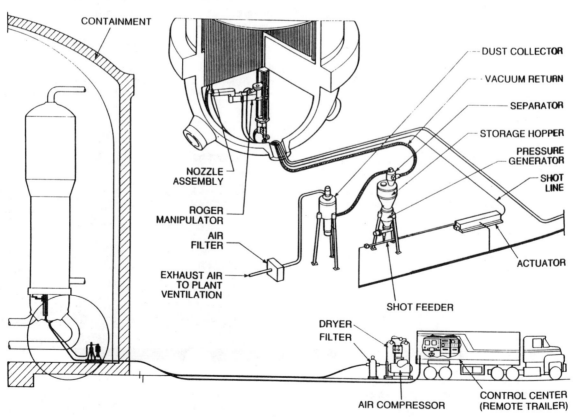

Fig. 6 - Schematic of Computer Controlled Shot Peening Equipment Used Inside of Containment at Nuclear Power Stations. [E.]

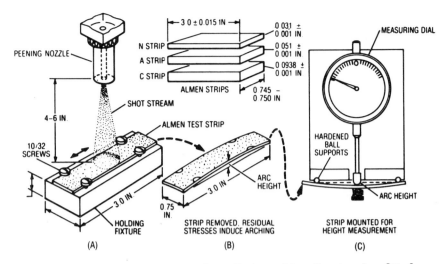

Fig. 7 - The Almen Strip System for Peening Intensity Control. [D.]

aircraft are shot peened to restore the compressive stresses lost through long service, high stresses, wear and corrosion. Coil springs for electronic equipment, so small that a thousand would fit in your hand, are peened to offset high service cycles. A vast array of shot peening machines and techniques are employed to address this kind of variety.

To be effective and repeatable, shot peening must be controlled. The media must be virtually free of broken, angular pieces that will introduce stress risers. The transferred energy from the shot to the part is controlled by the Almen system (Fig. 7). Coverage is best verified by fluorescent tracers that are removed by the impact of the shot in proportion to the distribution

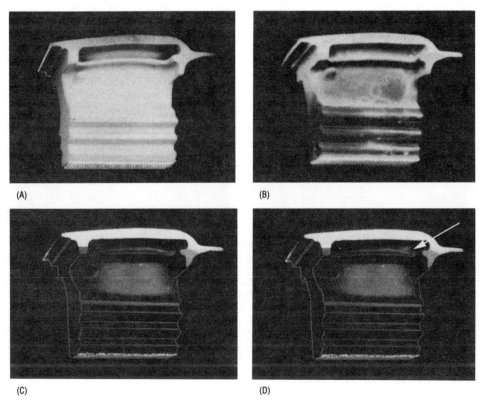

Fig. 8 - The Peenscan System. (A) Coated Unpeened (B) Peened 15 Seconds, Partial Coverage (C) Peened 60 Seconds, Full Coverage (D) Peened 60 Seconds, Improper Nozzle Angle in Cavity (as shown). [D.]

of the dimples (Fig. 8). Shot hardness relative to part hardness plays a very significant role in the parameters outlined above and in the magnitude of residual compressive stress that can be generated. Computers are now employed on shot peening machines to ensure the repeatability of these parameters as well as several others such as shot flow, air pressure, translation speed and cycle time.

DUAL SHOT PEENING

A recent study (Ref. 3) used x-ray diffraction techniques, per SAE J784a, to evaluate the effect of peening first with large shot to high intensity followed by a second peening with smaller media at low intensity. The material used in the test was SAE 8620, carburized and hardened to a maximum of HRC 62, a typical transmission gear steel. Table 2 shows the shot peening conditions that were studied. Figure 9 represents the residual stress profiles generated by the peening conditions. Several interesting observations can be extracted:

1. The baseline curve, for carburizing only with no shot peening, shows a remarkably neutral stress condition, at depth, with a small tensile stress at the surface. This tensile stress is attributed by the researchers to the presence of a shallow layer of upper transformation products and grain boundary oxidation. The tensile stress at the surface would have a debiting effect on fatigue life.

2. Peening such hard material with regular hardness (HRC 45-55) produced a relatively lower compressive stress profile (sample #84) when compared to sample #85, which was peened with hard shot (HRC 55-62) (Ref. 4). Interestingly, though, it took four times longer to achieve full coverage with the softer shot.

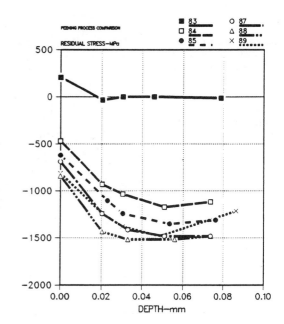

Fig. 9 - Comparison of X-Ray Diffraction Residual Stress Profiles for as Carburized and Shot Peened Samples Including Measurements Between the Surface and 0.02 mm. [F.]

3. Dual peened samples showed an increase in magnitude of residual compressive stress, at the surface and at depth and a corresponding increase in fatigue strength can be expected.

An excellent example of dual peening, with local interest, can be found in a recent Indianapolis 500, where eleven of the first twelve racing cars to "go the distance", did so on dual shot peened transmission gears (Ref. 5).

CONCLUSION

A good understanding of residual stresses and how they are generated can be of great value to engineers in many disciplines. Using compressive stresses as a beneficial tool often is the difference between success and failure of a mechanical system design.

Sample	Primary Treatment		Secondary Treatment	
	Shot Size(1)	Intensity(2)	Shot Size(1)	Intensity(2)
83	Not Shot Peened		-	-
84	230	0.46A (18A)	-	-
85	230H	0.56A (22A)	-	-
87	230H	0.56A (22A)	110H	0.15A (6A)
88	230H	0.56A (22A)	GP100	0.13A (5A)
89	230H	0.56A (22A)	GP100	0.13A (5N)
89(3)	230H	0.46A (18A)	GP100	0.13N (5N)
86(4)	230H	0.46A (18A)	GP100	0.13N (5N)

(1) Shot sizes conform to MIL-S-13165B, regular hardness cast steel shot is HRC 45-55, H denotes harder cast steel shot (HRC 55-62), GP denotes glass beads, MOH's hardness 5.5.
(2) Almen intensity A Strip, N Strip, in mm, English units in parenthesis.
(3) 0.29mm removed prior to peening by electropolishing.
(4) 0.13mm removed prior to peening by grinding, with coolant.

Table 2 Shot Peening Conditions Studied

REFERENCES

1. Almen, J. O., "Fatigue Failures are Tensile Failures", Product Engineering, Vol. 23, No. 9, pp. 101-124, 1951. Also discussed in "Impact", Fall 1989, Metal Improvement Company.

2. Fuchs, H. O., Editor, "Shot Peening Stress Profiles", Metal Improvement Company.

3. Ahmad, A. and Crouch, E. D., Jr., "Dual Shot Peening to Maximize Beneficial Residual Stresses in Carburized Steels", Carburizing Processing and Performance, Conference Proceedings, ASM International.

4. MIL-S-13165C, Military Specification, "Shot Peening of Metal Parts".

5. "Impact", Summer 1987, Metal Improvement Company.

REFERENCES TO FIGURES

[A.] Fuchs, H. O., "Regional Tensile Stress as a Measure of the Fatigue Strength of Notched Parts", Mechanical Behavior of Materials, Vol. II, pp. 478-488.

[B.] Friske, W. H., "Shot Peening to Prevent the Corrosion of Austenitic Stainless Steels", Atomics International Division, Rockwell International.

[C.] Koster, W. P., et al, "Surface Integrity of Machined Structural Components", Technical Report AFML-TR-70-11, March 1970.

[D.] "Shot Peening Applications", Seventh Edition, Metal Improvement Company.

[E.] Hellman, S. P., "Shot Peening to Prevent Stress-Corrosion Cracking of Nuclear Steam Generator Tubes", B & W Nuclear Service Company.

[F.] See Reference 3.